Engineering and Sustainable Community Development

Engineering and Sustainable Community Development

Juan Lucena, Jen Schneider, and Jon A. Leydens

www.morganclaypool.com

ISBN: 9781608450701   paperback
ISBN: 9781608450718   ebook

DOI 10.2200/S00247ED1V01Y201001ETS011

A Publication in the Morgan & Claypool Publishers series
SYNTHESIS LECTURES ON ENGINEERS, TECHNOLOGY, AND SOCIETY

Lecture #11
Series Editor: Caroline Baillie, *University of Western Australia*
Series ISSN
Synthesis Lectures on Engineers, Technology, and Society
Print 1933-3633   Electronic 1933-3461

Synthesis Lectures on Engineers, Technology, and Society

Editor

Caroline Baillie, *University of Western Australia*

The mission of this lecture series is to foster an understanding for engineers and scientists on the inclusive nature of their profession. The creation and proliferation of technologies needs to be inclusive as it has effects on all of humankind, regardless of national boundaries, socio-economic status, gender, race and ethnicity, or creed. The lectures will combine expertise in sociology, political economics, philosophy of science, history, engineering, engineering education, participatory research, development studies, sustainability, psychotherapy, policy studies, and epistemology. The lectures will be relevant to all engineers practicing in all parts of the world. Although written for practicing engineers and human resource trainers, it is expected that engineering, science and social science faculty in universities will find these publications an invaluable resource for students in the classroom and for further research. The goal of the series is to provide a platform for the publication of important and sometimes controversial lectures which will encourage discussion, reflection and further understanding.

The series editor will invite authors and encourage experts to recommend authors to write on a wide array of topics, focusing on the cause and effect relationships between engineers and technology, technologies and society and of society on technology and engineers. Topics will include, but are not limited to the following general areas; History of Engineering, Politics and the Engineer, Economics , Social Issues and Ethics, Women in Engineering, Creativity and Innovation, Knowledge Networks, Styles of Organization, Environmental Issues, Appropriate Technology.

Engineering and Sustainable Community Development
Juan Lucena, Jen Schneider, and Jon A. Leydens
2010

Engineering and Society: Working Towards Social Justice, Part I: Engineering and Society
Caroline Baillie and George Catalano
2009

Engineering and Society: Working Towards Social Justice, Part II: Decisions in the 21st Century
George Catalano and Caroline Baillie
2009

Engineering and Sustainable Community Development

Juan Lucena, Jen Schneider, and Jon A. Leydens
Colorado School of Mines

SYNTHESIS LECTURES ON ENGINEERS, TECHNOLOGY, AND SOCIETY #11

MORGAN & CLAYPOOL PUBLISHERS

ABSTRACT

This book, *Engineering and Sustainable Community Development*, presents an overview of engineering as it relates to humanitarian engineering, service learning engineering, or engineering for community development, often called sustainable community development (SCD). The topics covered include a history of engineers and development, the problems of using industry-based practices when designing for communities, how engineers can prepare to work with communities, and listening in community development. It also includes two case studies—one of engineers developing a windmill for a community in India, and a second of an engineer "mapping communities" in Honduras to empower people to use water effectively—and student perspectives and experiences on one curricular model dealing with community development.

KEYWORDS

humanitarian engineering, sustainable engineering, service learning engineering, engineering for community development, sustainable community development (SCD), Engineers Without Borders (EWB), appropriate technology, social justice, international engineering

Contents

Preface

We wrote this book because we care deeply about engineers and engineering education. One author of this book (Lucena) holds degrees in mechanical and aeronautical engineering, and the three of us have devoted a combined 40 years of our lives to educating engineers. We read and have published in several engineering education journals, and two of us have co-edited a journal on engineering studies. Also, we have conducted numerous interviews with students and faculty involved in engineering design. At our own university, we participated in the creation and implementation of a minor in Humanitarian Engineering, which from its inception placed senior design as its most important course. These experiences have shaped our view of engineering education, particularly as it relates to sustainable community development (SCD). In many ways, these experiences have also inspired us to write this book.

As we talk with students and faculty and closely study syllabi, engineering education journals, and conference proceedings, we have also come to realize that many engineering faculty and students involved in SCD activities hold problematic assumptions and follow industry-inspired practices and methods when trying to work with communities. These experiences and realizations have motivated us to write *Engineering and Sustainable Community Development* with the goal of helping all of us involved in engineering education to critically reflect on our practices, question their appropriateness for SCD, and begin transforming them to meaningfully include community.

We want to thank all the reviewers who took the time to read our chapter drafts and found constructive ways to provide us with honest and valuable feedback: Chris Byrne, Nalini Chhetri, Rebekah Green, Kelly Jones, Nina Miller, Dean Nieusma, Marie Paretti, and Cameron Turner. We want to acknowledge in particular our colleague David Frossard, who has walked with us many of the paths that led to this book, particularly the teaching of the course "Engineering and Sustainable Community Development" and the revisions of many of the book's chapters. Many people were involved in the creation and delivery of this course, especially a core group of faculty who met regularly for three years and participated in its piloting: Tina Gianquitto, Carl Mitcham, Barbara Moskal, Junko Munakata-Marr, Dave Munoz, Marcelo Simoes, and Jay Straker. Gustavo Esteva and Anu Ramaswami, colleagues who visit our course regularly and give us inspiration and insights, greatly influenced us in making community central in engineering projects and activities. Several undergraduate and graduate students were active participants in improving our course and our understanding of SCD.

We want to express our most sincere thanks and appreciation to those at the National Science Foundation who created and supported the program Ethics Education in Science and Engineering (EESE), in particular Rachelle Hollander and Sue Kemnitzer. EESE funded the grant that made this book possible (NSF Grant # EEC-0529777).

We also want to thank Caroline Baillie, our colleague and series editor, and Joel Claypool, our publisher, for giving us the opportunity to be part of the series *Synthesis Lectures on Engineers, Technology and Society*, which holds the promise to help us rethink what engineering is for, and how we should teach the engineers of tomorrow.

Our final thank you goes to our spouses and children, who gave up many weekends, playtimes, and hiking trips so their parents could finish this manuscript.

We dedicate this book to the next generation of engineers who hold the humility, perseverance, and courage to join in bringing about sustainable community development.

Juan Lucena, Jen Schneider, and Jon A. Leydens
February 2010

CHAPTER 1

Introduction

1.1   ENGINEERS AS PROBLEM SOLVERS

How often have you seen yourself as a problem solver? How many times have others associated you with solving problem because you are an engineer? Have you found yourself wanting to apply your problem solving knowledge and skills to the problems of underserved communities? Have your professors challenged you to apply engineering to solve problems of communities in need? If you answered yes to at least one of these questions, this book is for you.

Historically, engineers have been identified primarily as problem-solvers (Koen, B., 2003; National Academy of Engineering, 2004; Downey, G., 2005). However, we argue in this book that this identity could be highly problematic for sustainable community development (SCD) projects if the dominant approaches to problem solving are the Engineering Problem Solving approach (EPS) or industry-based design learned throughout engineering curricula. Downey and Lucena (2006) have argued that

> the technical five-step engineering method (Given, Find, Diagram, Equations, Solution) that is still taught regularly in engineering science courses at the core of engineering curricula includes no mechanism for addressing the routine non-technical problem of working with people who draw boundaries around problems in different manners …students who complete hundreds of problem sets on graded homeworks and exams are simultaneously receiving intensive training in dividing the world of problem solvers into two parts, those who draw boundaries around problems appropriately and those who do not. The first group becomes capable of being "right," while the second is, by implication, "wrong." Quality students emerge from engineering curricula knowing that engineering problems have either right or wrong answers, that the chief metric of ability is the frequency that one is right, and that difference is usually a sign of error. In the process, they have acquired solid grounds, seemingly mathematical, not to trust the perspectives of co-workers who define problems differently. In other words, learning the five-step engineering method appears to make a diversity of viewpoints suspect by definition (Downey and Lucena, 2006).

The history of engineers and development, as we will see in Chapter 2, suggests that the engineer as problem solver might fit well with the technical dimensions of development projects. But how about its non-technical dimensions? How might engineering problem solving and design approaches as currently taught in your curriculum be at odds with the socio-economic, political and cultural dimensions of development, particularly those associated with the communities that development is supposed to serve?

1.2   ENGINEERS' BELIEFS ABOUT COMMUNITY DEVELOPMENT

Likely, as an engineer, you have great faith in the power of technology to solve human problems. Furthermore, you probably believe that technology enhances people's quality of life and makes economies grow. Perhaps, these beliefs drive you, your peers, and your faculty to develop technological solutions for communities in need in spite of the economic, cultural and geographical distance that exists between your locality and the communities.

No one would question that engineers make technology happen. In large part, the power that engineers have over technological development lies in their knowledge, skills, attitudes, and beliefs towards technology. Yet some of these attributes are highly problematic when it comes to developing technological solutions for communities, for multiple and complex reasons. We will explore some of these throughout the book. Here we would like to call your attention to a number of important beliefs that perhaps explain why the relationship between engineers and community development continues to be highly problematic after more than 60 years and billions of dollars spent in international development:

- Many engineers continue to believe in the main premises of *development and modernization*, particularly, a) that a socially engineered order, informed by science and realized through technology, will bring progress (e.g., that a more efficient distribution of water will lead to increases in quality of life in a community); and b) that technological development will lead to economic growth and then to increasing human satisfaction and welfare. In Chapter 2, we will explore some historical connections between engineering and modernization and show how this ideology has permeated engineering for development work since the 1950s. In Chapter 4, we will further elaborate how engineers' belief in modernization and development is highly problematic for community development projects.

- Most engineers continue to uncritically believe in the power of *technology to transform society* yet hold on to the assumption that technological development happens independently from society, culture, or politics (held by most Americans, not just engineers, this belief is also known in the literature as technological determinism. See Smith et al., 1994). Note, for example, how often engineers describe the technical vs. nontechnical dimensions of a senior design project as if these two worlds were actually separate, only coming into contact during implementation and use of the design in question. This belief is highly problematic in community development, for it places engineers as experts in control of technological development and community members as passive receivers of a technology already developed. It also leads engineers to assume that a community's social, cultural, and political dimensions have very little to contribute to technological development. Throughout the book, we will explore the problems associated with this belief, especially in design for community activities in Chapter 3 and as a potential limitation to effective listening in Chapter 5.

- In spite of decades of ethnographic work, mainly from cultural anthropology and development studies, that shows the complexity and heterogeneity of communities, most engineers in development assume that *communities are homogeneous* entities with one voice and can be treated as a "client" or "customer" in a for-profit relationship or in design for industry projects. We will explore the heterogeneous complexity of communities through two mini case studies in Chapter 4, propose a more effective way for listening to the diversity of perspectives in Chapter 5, and show how engineers have dealt with this diversity in specific community development activities in Chapters 6 and 7.

- In spite of extensive evidence on the complexity of technology transfer across cultural contexts, many engineers continue to believe in the *universality* of technological applications while undervaluing or ignoring the differences that exist among local, regional, or national contexts in the design, appropriation, implementation, diffusion, and use of technologies. We will explore how two engineers questioned this belief in their practice and empowered communities to define the technologies that they wanted in Chapters 6 and 7.

- In spite of the fact that cross-cultural engineering teams are becoming the norm in private corporations, many engineers ignore the complexities in working at the nexus of international development organizations, NGOs, and community projects where cultural and political nuances are even more complicated than in the private sector. Unfortunately, most engineers ignore the difficulties of listening to different perspectives across different cultures, organizations and/or within communities. We will explore the complexities of listening in community development contexts in Chapter 5.

So, is engineering for sustainable community development doomed to fail because of the problematic involvement of engineers in development and their uncritical attitude towards the assumptions and beliefs listed above? The answer may be "yes" if engineers continue to uncritically hold on to these beliefs, ignore the history of their involvement in development, and take community for granted. The answer might be "no" if those of us committed to sustainable community development engage in a serious re-examination of our history, practices, education, assumptions, and more. This book is the beginning of such a re-examination.

1.3   ENGINEERS, DEVELOPMENT, AND COMMUNITY

For some years now, through our research and teaching, we have been encountering engineers doing what might be called "engineering for development." That is, engineering work in parts of the world that many in the US consider "third," "developing," "underdeveloped," "poor," or "underserved." Many of these engineers have made significant contributions to the field of engineering and sustainable community development (ESCD), and there is much to be learned from their stories.

One of these engineers, who asked for anonymity, has more than four decades of experience in international development. He told us how his perception of development work changed over time. We quote him at length here because we believe his story is important and illustrative:

> ...all the millions and millions of dollars that the World Bank had given to these [developing] countries or loaned to these countries ...wasn't helping the poverty-stricken people in the villages. In fact, poverty was increasing, not decreasing. So I kept thinking, well, what is it that needs to be done differently? Then one of our graduate students...in civil engineering...had gone to work for...[a] very fine engineering firm mainly on the East Coast. And [that firm] put in a proposal and [US]AID [the United States Agency for International Development] funded them to put in water supply and sanitation [systems] in many villages because half of the children who were dying before the age of six [were dying] because of water-related diseases....
>
> [U]sually [these children would] have dysentery, and it would make them so weak that when measles came along, they couldn't fight off the measles. So that made the United Nations declare the decade of the 1980's as the [International Drinking Water Supply and Sanitation Decade]. Well, this fellow...was always telling me all the wonderful things that they had done, hundreds and hundreds of villages that now had a clean water supply and sanitation [system]. And that ended, of course, in 1990, the decade of the 1980's. Five years later, in 1995, they conducted a survey, and found that only 30 percent of those [systems] were still in operation. And then in the year 2000, they did a survey, and found that only 12 percent were still in operation.

For this engineer, engineering for development was inspiring, yet it was also frustrating because engineering projects in the "developing" world did not often last and could not be considered long-term or "sustainable" solutions to any of the problems they were intended to solve.

Another engineer we spoke to reiterated similar concerns. From the 1990s to the present, Elena Rojas worked, hands-on, with water infrastructure projects in Honduras. There, she learned that technical engineering aspects were the easiest problems for her to solve. Contrary to her previous beliefs, what proved to be much more difficult was to figure out how "technical" projects fit into particular communities, who supported them, and who would be using them:

> You start understanding...that you can be a very good technical engineer, and...that's not really a challenge at all.... But...once those projects are implemented, what is the key issue to make sure that they will last the time you have planned they should last? So that's something that you as a technical person cannot solve if you do not take into account the *people* that are going to be taking care of or using those systems.

The experiences of the two engineers mentioned above point us to three main themes central to the work of this book: "community," "communication," and "help." So what do development engineers need to know about communities, about how to communicate with them, and about their own desire to help? Community development practitioners have argued that for a development project to be truly sustainable, among other variables the local context of its implementation—community–must be seriously considered. After all, it is people in specific localities that can make or break a project and ensure (or disrupt) its long-term sustainability. But this can be difficult, especially when development funds are "mainly invested in the building of new facilities, not in building the capacities of rural communities" (Gomez et al., 2008, p. 231). In other words, development work tends to be oriented toward completing particular, discrete projects, regardless of whether there has been an understanding of the community that the project is supposed to serve. "Therefore, the lack of institutional support and financial aid to empower communities affects the real sustainability of systems. The process of technology transfer will probably continue for some time, but its success is far from certain because of the weak role played by the community and institutions in the appropriation of new technologies in the long term" (Gomez et al., 2008). Hence, we propose in this book, as shown in Figure 1.1, that *community needs to be central* to development projects, particularly to the engineering dimensions of these, if we hope for these projects to be sustainable in the long run and to increase the self-determination of the communities that they are intended to serve.

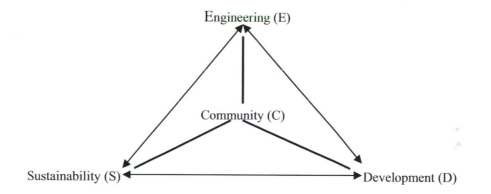

Figure 1.1: A community-centric model of ESCD implies that development (D) and engineering (E) should be for and about community (C) and that sustainability (S) will not happen without community's self-determination and ownership of projects.

Yet even the idea of making "community" central to engineering for development is more complex than it sounds. It seems obvious to state that if we want to help a community, we must first understand it by coming to know, value, and communicate with its members. But we have found that even the catalyst for beginning such projects—our initial desire to help—can be problematic and should be critically examined. Gustavo Esteva, a grassroots activist and intellectual in Oaxaca, Mexico, is particularly wary of this drive to help. When Americans or others offer "help," he urges, "Don't come [to my community] to help! Come to listen, to find out if our struggles are your struggles. Then and only then, we can sit and discuss how, if at all, we can work together" (personal communication).

Esteva's challenge is a challenge to all of us in engineering who are committed to development work. It should raise many questions for us and encourage our self-reflection. It is the primary goal of this book to help engineers involved in any aspect of development to understand the challenges of working with communities, to question their desire to "help," and to re-imagine themselves not just as "problem solvers" but also as listeners and facilitators who enhance communities' capacities to chart the course of their own future.

For development projects in engineering education to evolve into true sustainable community development projects, engineers need to incorporate communities' histories, voices, concerns, conflicts, knowledges, desires, and struggles by learning how to listen and recognize value in the perspectives of others, including non-experts. This requires rethinking the preparation students receive (particularly in engineering design) to participate in such projects, to potentially include courses in development studies, fieldwork methods, regional history, or others. We suggest that the long-term success and ethical integrity of these projects will largely depend not only on the safety and reliability of their technical components but on the critical involvement of community members and engineers' ability to critically question their own motives to help.

Placing community at the center of development projects and initiatives remains a largely unmet challenge in engineering education; it is a challenge that hopefully will not become invisible, as has happened in engineering for development work in the last five decades (see Chapter 2) and continues to happen in many "design for community" activities (see Chapter 3), but one that will come to define the core of engineering education. We address this challenge in more detail in Chapter 4, propose ways for listening to communities in Chapter 5, show two case studies of community engagement by engineers in Chapters 6 and 7, and propose a curricular alternative in Chapter 8.

1.4   WHAT MAKES THIS BOOK DIFFERENT?

Our approach to Engineering and Sustainable Community Development (ESCD) differs from other work on engineering and sustainability (Prendergast, J., 1993; Manion, M., 2002; Mulder, K., 2006) because of our focus on community, communication, and engineers' desire to "help" (which turns out to be more problematic than seen at first glance). Much of the focus in engineering for sustainability is on evaluating technical reliability or environmental impact (e.g., carbon footprint or the connection between technosphere and biosphere) of a particular system, process, or project; community often appears as a marginal add-on and poorly understood concept to such concerns. We think, however, that critical approaches to community, communication, and helping should be the *cornerstones* of engineering for development work—as foundations of sustainability—rather than afterthoughts. Here is a list of questions that have emerged out of our explorations into the relationships among engineering (E), sustainability (S), community (C), and development (D). These questions have guided the content and organization of this book. We offer them as a list of questions that we hope you, as future practitioner of engineering for development, will ask:

Guiding Questions

1. What can we learn from the experiences of engineers who have done development work in the last five decades?

2. Why do many development projects—even those that are small-scale, "appropriate," and otherwise "sustainable"—fail? What can we learn from those failures? Who, if anyone, should be held accountable? What are the characteristics of successful approaches to sustainable community development? How can engineers learn these?

3. What is problematic about engineering problem solving or design for industry approaches when dealing with projects for sustainable community development?

4. What are the possible motivations for doing engineering for development? Do these motivations matter for the outcome of projects? If so, who benefits and who does not from these motivations?

5. If engineers aren't supposed to be "helping" a community, then what are they supposed to do? How can they find other motivations to work with community?

6. What do we mean when we speak of "communities"? What are important mindsets and strategies for communicating and working with communities? How do you listen to a community?

7. How can you understand your own struggles as a development engineer? How can you understand how community struggles intersect with your own? What if there is no intersection?

8. What kind of alternatives do we have in engineering education to provide future engineers with knowledge, mindsets, and strategies to work with communities?

This book addresses these questions, some in more detail than others, and provides case studies of engineering practice and education to show how engineers, in their practice and in the classroom, have grappled with them. Our main claim is that if engineers are committed to the *sustainability* of engineering development projects, and to *community self-determination* through those projects, they must think critically about their motivations, approaches, and relationships to those communities. These critical reflections must impact their practices. In the long run, neither sustainability nor development will happen without placing community as the focal point. As John Blewitt, director of the master's degree program in Sustainable Development at University of Exeter (UK), reminds us, "ecological sustainability can only be attained through social learning, community empowerment, participation, and a commitment to global justice" (Blewitt, J., 2008). Engineers need to understand the complex connections among sustainability, community, and development, since historically engineers have been important agents in development work; more recently they seem to be increasingly committed to sustainability but, for the most part, have taken community for granted.

This book is not a step-by-step handbook on how to do engineering for SCD. Rather, it is an invitation for critical reflection, both individually and professionally. We hope to evaluate the extent to which you are committed to the self-determination of communities, to promoting sustainability in your actions as an engineer, and to putting the book's ideas into practice.

1.5   WHO IS THIS BOOK FOR?

Where are engineers for development being educated nowadays? Where are other students like you learning how to apply engineering knowledge and methods to community development problems? An informal survey of engineering programs in the United States clearly shows that a growing number of universities offer classes, initiatives, programs, or degrees in engineering and sustainable development, community service, service learning, and/or humanitarian engineering. It's not possible to include an exhaustive list here, but some examples of these types of programs include:

- Community Assessment of Renewable Energy and Sustainability (CARES) at University of California, Berkeley

- Design for the Poorest 80% course at Michigan University

- Engineering and Humanity Institute at Southern Methodist University

- Engineering for Developing Communities (EDC) Program, University of Colorado

- Engineering Projects in Community Service (EPICS) Program, founded at Purdue University and now a national network of programs

- Engineers for Community Service, Ohio State University

- Engineers in Technical and Humanitarian Opportunities of Service (ETHOS) at Iowa State, the University of Dayton, Humboldt State University, Colorado State University, and the University of Illinois

- Entrepreneurial Design for Extreme Affordability at Stanford University

- Humanitarian Engineering, Colorado School of Mines

- Humanitarian Engineering & Community Engagement Program at Penn State University

- Humanitarian Engineering Leadership Projects (HELP) at Dartmouth College

- Peace Corps Master's International Program, including a master's degree in engineering, Michigan Tech

- Master's International Peace Corps Program, including a master's degree in engineering, University of South Florida

Although these programs are conducted under a number of auspices and with varying objectives, they seem to share one thing in common: an expressed desire to use engineering to "help" communities "in need." We hope that students, faculty, and administrators teaching and working in these programs will find this book useful and provocative. This book also hopes to inform those involved in accreditation, funding administration, and policymaking of engineering education who have begun to face questions about the usefulness, effectiveness, and cost of these programs as interest and demand for these programs continues to increase. This book also complements similar efforts to understand connections among engineering, local and global communities (Baillie, C., 2006), engineers, poverty and peace (Catalano, G., 2006), and engineers and social justice (Riley, D., 2008).

We hope that after reading this book, all readers will develop an awareness of the importance of integrating community and sustainability efforts in engineering education and develop and implement new criteria for accreditation, funding, and support. Because of our own location as scholars working in the United States, we are most familiar with programs here, and so this book is skewed toward US-based initiatives. However, we hope that many ideas from this book can be extended to programs outside the US where they are also flourishing.

Perhaps you have noted that, at this historical moment, more and more engineering jobs seem to be directly or indirectly related to SCD activities. Maybe you have begun exploring these new career trajectories outside of mainstream engineering employment. You and your peers need to be aware of this trend, its possible causes and implications for your future careers:

- **Have you noticed that more engineering jobs are directly or indirectly linked to sustainability or SCD-related areas?** Given the scope and number of environmental challenges we face, from climate change to fresh-water depletion, engineers are being called on in greater numbers to work in fields related to sustainability. If you conduct an Internet search for the phrases "engineering jobs" and "sustainable development," you will see that the term "sustainable" is now part of engineering titles of "sustainable design engineer," or "sustainable materials engineer," and the title "sustainable development engineer" is actually used in a number of companies. Descriptions such as "civil engineer" jobs with the US Navy now include duties such as "directing overseas construction of critical humanitarian importance."

- **Why might mid-career engineers be seeking employment or volunteer opportunities in non-traditional organizations involved in SCD activities?** Engineers who might feel "stuck" in, or who have been laid off from, companies that comprise the military-industrial complex, the auto industry, or the Internet sector that bubbled in the 1990s might be thinking about a career change towards sustainability- or community-development related careers. Many of those who cannot leave their jobs for financial security reasons are volunteering in SCD-related areas in higher numbers. Note the increasing number of volunteering opportunities in com-

Figure 1.2: Nick Edwards and fellow Dartmouth engineers built and installed a solar-powered water pump in Nyamilu, Kenya and designed and constructed a distribution system for the water source during a 2007 trip to the small village in Africa.
(Source: `http://www.flickr.com/photos/thayerschool/2782139824/` Credit: Darmouth Humanitarian Engineering Leadership Projects (HELP).)

munity service and humanitarian relief through Engineers Without Borders and engineering professional societies. Many private companies endorse these volunteer activities as a way to improve their image for corporate responsibility. In 2009, President Barack Obama signed into law the Volunteers for Prosperity (VfP) program in the US Agency for International Development (USAID). "The VfP program aims to promote short- and long-term international volunteer service by skilled American professionals to addressing the needs of those living in the poorest areas of the world" (`www.usaid.gov/about_usaid/presidential_initiative/vfp.html`). We expect that many engineers will participate in this program.

• **Why are new job opportunities emerging in communities throughout the US as economic recovery projects scale up?** As federal funding for renewable energy, "green jobs," and community-based initiatives continues to increase, more and more engineers will face the challenges of meaningfully engaging local communities in sustainability efforts.

This book hopes to serve future engineers in the US and elsewhere, who will be facing both exciting opportunities and some daunting challenges when working with communities.

1.6   BRIEF OUTLINE OF THE BOOK

- Chapter 2 outlines a history of engineers in development from their involvement in the expansion of empires in the 19th century to their most recent dealings with sustainable development. This historical trajectory shows how the key concepts of "community" and 'sustainability' remained invisible to engineers for most of the 20th century and how their recent emergence has been a significant challenge to engineering education and practice.

- Chapter 3 is an anatomy of a senior design project and course as traditionally taught in engineering education. It shows how adopting assumptions, practices, and models from "design for industry" to "design for community," leads engineering educators, and students to *disable the self-determination* of the communities they want to serve, making them practically invisible.

- Chapter 4 addresses how engineers have tried to engage *community*, the strengths and limitations of these approaches, and the conceptual and methodological challenges posed to engineers by community. Only by taking communities seriously ETH engineers can actually begin to serve them.

- Chapter 5 focuses on listening as an important dimension of community engagement. The chapter also analyzes how engineering curricula may inhibit listening, proposes that listening is a key missing dimension in engineering education, and argues that contextual *listening* is a prerequisite to understanding, valuing and respecting the diversity of perspectives involved in SCD activities. Understanding these perspectives is, in turn, critical to long-term SCD project success.

- Chapters 6 and 7 are two case studies of engineers' involvement with communities where listening to the community led to reconfiguring the initial project, challenged engineers' identities, and helped engineers reconsider their preconceived assumptions about community. Chapter 6 involves an engineering professor developing a windmill for a small community in India, and Chapter 7 involves a practicing engineer "mapping communities" in Honduras to empower people to use water safely and effectively.

- Chapter 8 describes a curricular experiment to help students and faculty understand the challenges and opportunities involved in the development and teaching of an interdisciplinary course in *Engineering and Sustainable Community Development (ESCD)*.

- The book concludes with a realization that the relationship between engineers and community is only one among a myriad of relationships among governments, institutions, actors, and local communities involved in ESCD projects. We invite readers to reflect and prepare themselves for the complexity involved in ESCD work and to consider *social justice* as the next challenge in their engineering activities.

The chapters are written in such a way that they can be excerpted for use in classes, but also could be read independently by readers interested only in particular areas of focus. At the same time, to get the most out of this book, it should be ideally read in its entirety, as all the chapters are interconnected and intentionally reinforcing, thematically and philosophically.

REFERENCES

Baillie, C. (2006). *Engineers within a local and global society*, Morgan & Claypool. 9

Blewitt, J. (2008). *Community, empowerment and sustainable development.* Totnes, Green Books. 7

Catalano, G. D. (2006). *Engineering ethics: Peace, justice, and the earth*, Morgan & Claypool. 9

Downey, G. and J. Lucena (2006). Are Globalization, Diversity, and Leadership Variations of the Same Problem? Moving Problem Definition to the Core. *ASEE Annual Conference and Exposition.* Chicago, IL, ASEE. 1

Downey, G. L. (2005). ""Are Engineers Losing Control of Technology?: From "Problem Solving" to "Problem Definition and Solution" in Engineering Education." *Chemical Engineering Research and Design* **83**(A8): 1–12. 1

Koen, B. V. (2003). *Discussion of the method : conducting the engineer's approach to problem solving.* New York, Oxford University Press. 1

Manion, M. (2002). "Ethics, engineering, and sustainable development." *IEEE Technology and Society Magazine* **21**(3): 39–48. 6

Mulder, K. (2006). *Sustainable Development for Engineers: A Handbook and Resource Guide.* Sheffield, UK, Greenleaf Publishing. 6

National Academy of Engineering (2004). *The Engineer of 2020: Visions of Engineering in the New Century.* Washington, DC, The National Academies Press. 1

Prendergast, J. (1993). "Engineering sustainable development." *Civil Engineering, ASCE* **63**(10): 39–42. 6

Riley, D. (2008). *Engineering and social justice*, Morgan & Claypool. 9

Smith, M. R. and L. Marx, Eds. (1994). *Does Technology Drive History? The Dilemma of Technological Determinism.* Cambridge, Mass., MIT Press. 2

Trejos Gomez, Garcia-Zamor, et al. (2008). "Domestic Wastewater Management in a Rural Community in Colombia." *Comparative Technology Transfer and Society* **6**(3): 212–35. 5

C H A P T E R   2

Engineers and Development: From Empires to Sustainable Development[1]

How did engineers get involved in development? How have engineers been engaged in imperial, national, international, and sustainable development? How have historical ideological, and institutional factors influenced the way engineers engage with the groups of peoples (tribes, communities, villages, etc.) that they are supposed to serve? To what extent might this history constrain engineers' ability to effectively define problems and implement solutions for sustainable development? The answers to these questions will help you envision future possibilities and hidden limitations for individual and professional involvement in sustainable community development (SCD) and humanitarian engineering in more realistic, critical, and humane ways.

This chapter traces episodes of the history of engineers' involvement in development, from 18th century colonial development to 21st century sustainable community development. As you travel through the chapter, take the time to pause and answer the critical questions and exercises posed along the way. These are intended to elicit reflection on how much the history of engineers' involvement with development might continue to shape the ways in which you engage community development or humanitarian engineering today.

2.1 ENGINEERS AND THE DEVELOPMENT OF EMPIRES (18TH AND 19TH CENTURIES)

The emergence of engineers, engineering practice, and engineering education has a close connection to the development of countries (Downey and Lucena, 2004; Lucena, J., 2009a,b). When countries developed as empires and colonies during the 18th and 19th centuries, engineers worked both for the internal organization and expansion of the empires and in the colonies as agents of imperial development (Mrazek, R., 2002).

---

[1]Some parts of this chapter originally appeared in Lucena and Schneider, "Engineers, development, and engineering education: From national to sustainable community development," European Journal of Engineering Education, 33:3 (June), pp. 247–257.

Key Terms

Empires: Countries like Britain, France, Portugal, Spain, and the US that from the 18th to 20th centuries expanded their influence around the word by conquering and colonizing other countries or territories, often for the extraction of natural resources and human labor and/or the creation of markets.

Colonies: Countries like Brazil, Egypt, India, Mexico, and the US that were governed and, in most cases, exploited by empires.

For example, Spanish engineers, with significant influence from French military engineers, built military and civil infrastructures in Spanish colonies in the Americas (Galvez, A., 1996). French engineers worked in Egypt in the construction of the Suez canal (See Figure 2.1) (Regnier and Abdelnour, 1989; Moore, C., 1994). Later, British engineers worked in Egypt (Mitchell, T., 1988) and India (Cuddy and Mansell, 1994) to improve transportation and irrigation infrastructures that would facilitate imperial control and the extraction of natural resources (Headrick, D., 1981, 1988). German and British engineers worked for their imperial companies in mining extraction in Brazil (Eakin, M., 2002). Although working in different parts of the world and under different relationships between empire and colonies, these engineers shared a primary concern: *permanent transformation*, i.e., the attempt to transform nature into a predictable and lasting machine that could be controlled and would last to ensure their imperial patrons a return on investment and display superiority over indigenous people.

How did engineers, and the imperial governments that hired them, perceive and affect communities during these developments? Answers to this question yield insight into how engineers in some colonial contexts conceptualized and interfaced with communities. In most cases, communities became sources of forced labor to extract natural resources necessary for the construction of imperial projects. Quite often, natives were viewed as potential imperial subjects to be organized in ways that made it possible to tax them, convert them to Christianity (or the dominant religion of the empire) and often force them into labor. By design or by default, engineers working for empires were involved in the political re-organization of indigenous populations and their communities, by surveying and drafting maps of the colonies, building roads and bridges connecting city and country, and ports to facilitate the extraction of wealth from colony to empire (Lucena, J., 2009a,b). In short, the political and economic interests of empires over colonies, and the socio-economic and ethnic backgrounds of the engineers (most of whom were paid imperial employees born and educated in Europe, who generally considered themselves superior to colonial natives) dictated this kind of exploitative relationship between engineers and communities.

Figure 2.1: Opening of the Suez Canal in 1869. This major engineering project, authorized by the Ottoman governor Sa'id of Egypt, built by a French company and later used by the British empire, clearly represents engineering for the development of Empires.
(Source: `http://www.canalmuseum.com/documents/panamacanalhistory023.htm`   Credit: canalmuseum.com).

Critical Questions

When envisioning your participation in a community development or humanitarian engineering project or initiative, how do you see yourself in relationship to the community with which you are working (technologically, culturally, spiritually, in terms of your respective humanity, etc.)? As superior? Equal? Inferior? Be as honest as you can. What might be the justification for your sense of superiority, equality, or inferiority?

2.2 ENGINEERS AND NATIONAL DEVELOPMENT (19TH TO 20TH CENTURIES)

As independent republics began to emerge in the world scene, as happened first in the American continent beginning in the late 18th and early 19th centuries, engineers from these new nations became preoccupied with mapping the territory and natural resources of newly sovereign countries and building national infrastructures. Now born, and in some cases educated, in the former colonies, engineers adopted national identities and became preoccupied with developing their new countries. Through new infrastructures—mainly roads, bridges, railroads, canals, and ports—engineers helped connect widely dispersed and diverse populations into a national whole and integrate their productive capacity for national and international markets. New engineering schools emerged with these developments. For example, in 1820 the US government began training military engineers at West Point to provide the new republic with the necessary infrastructure that would protect it from future European invasions (Walker, P., 1981; Grayson, L., 1993; Smith, M., 2008). In the 1840s, the US Corps of Engineers used slaves to construct coastal defenses in the Florida Keys (Smith, M., 2008). Right after independence in 1821, engineers from Mexico's Colegio Nacional de Mineria began mapping their territory and building a civil infrastructure that would serve the newly independent country (Lucena, J., 2009a). In 1847 and with similar purposes in mind, engineers from Colombia's newly created Colegio Militar developed the first national system of roads and built the national capitol building (Safford, F., 1976, Ch. 7). Immediately after the creation of the Brazilian Republic (1889), military engineers from the Escola Politecnica de Rio connected the hinterlands of the Brazilian Amazon with the rest of the country through an extensive telegraph network (See Figure 2.2) (Diacon, T., 2004).

Figure 2.2: During his expeditions to build an extensive telegraph network across the Brazilian territory to unite Brazil, military engineer Candido Rondon da Silva tried to persuade indigenous groups in the Amazon to embrace the Brazilian nation.
(Source: http://www.vidaslusofonas.pt/candido_rondon2.htm Credit: Museu do Indio, Rio de Janeiro. Permission Pending).

Quite often, foreign engineers were invited to work alongside engineers from the newly independent countries when these did not have the financial capital, in-house experience, engineering education institutions, or machinery to build infrastructure projects. For example, French engineers were invited by the US government to develop engineering curricula in West Point Military Academy and build and supervise road construction (Walker, P., 1981). Francisco Cisneros, a Cuban American engineer educated at Rensselaer Polytechnic Institute (founded in 1824), was invited to Colombia to build the railroad and fluvial transportation systems (Horna, H., 1992). US and Canadian engineers were invited to Sao Paulo, Brazil, to develop the automobile industry and construct urban electric rail transportation (Telles, P., 1993). Yet neither local nor foreign engineers conceived these projects with environmental sustainability or community development, as we understand those terms today, in mind. Rather, consonant with the values of the day, nature and community were to be controlled and exploited for nation building.

Key Terms

Positivism: The belief that we can know and understand the world only through empirical, scientific observations and testing. For positivists, scientific reasoning is a superior, universal, and objective way to understand the world, while other forms of seeing and being in the world are considered inferior, superstitious, local, and ultimately unprovable.

Spencerism: A view of the evolution of society, first developed by English philosopher Herbert Spencer (1820-1903), in which society is considered an "organism" that evolves from simpler states to more complex ones according to the universal law of evolution. For the organism to survive and evolve, every part of society serves a function under an established hierarchy controlled by the State. Under this view, professions such as engineering play key roles in organizing important activities for the functioning, survival and evolution of the organism (e.g., infrastructure, industry) while marginal groups (poor, illiterate, orphans, etc.) and native communities are considered detrimental to the organism.

Social Darwinism: A view of human society rooted in Darwin's notion of survival of the fittest used to justify the superiority and authority of one group of people (usually whites, rich, educated) over other groups of people (usually non-whites, poor, and uneducated). Note that Darwin did not intend his notion of survival of the fittest to be applied to human societies.

During the late 19th and early 20th centuries, engineers in many parts of the former European colonies were heavily influenced by the ideologies of positivism and Spencerism, defined briefly above (Nachman, R., 1977). According to these ideologies, the purpose of the State was to establish *order* among a country's population to achieve *progress*. Spencerism was used to justify the actions of the State (and, by implication, the actions of engineers). An example from Mexico illustrates how engineers were involved in this "ordering" of society. According to one historian, Mexican engineers hired by President Porfirio Diaz (1876-1911) were part of a "brain trust of Positivists and Social Darwinists." This group of men believed that "government policy should be carried out according to the rules of 'science'" (Haber, S., 1989, p. 23).

In other words, Mexican positivists argued that, like an organism, society has many parts that should perform specific functions. In a certain sense, society could be viewed as a system that needed to be engineered for maximum efficiency. According to these thinkers, for a country like late 19th century Mexico to achieve order, the State had to instruct educational institutions to "educate" all people—regardless of ethnic and linguistic differences found among millions of indigenous peoples organized in hundreds of communities—into national citizens who would think and act alike. Meanwhile, some adult citizens would be transformed into professionals by professional and technical educational institutions (Bazant, M., 1984, 2002). Once educated on how to execute superior functions (e.g., build transportation infrastructure, industry), engineers, like other professionals, could contribute to the survival and evolution of society. Only through this level of *order* would a society (organism) ensure its survival and *progress*. Although not all countries adopted Spencerism as an ideology to organize society and justify the role of engineers, other examples can be found in Brazil during the first government of President Getulio Vargas (1930-1945) (Williams, D., 2001) and in Colombia during the last two decades of the 19th century (Henderson, J., 2001).

As you might imagine, under the ideologies of Spencerism and positivism, engineers and communities often clashed. Engineers were frequently in a position to *socially engineer* communities for the purposes of order and national progress, for example, by relocating them or connecting them in different ways to other parts of the country. For instance, Candido Rondon Da Silva was one of Brazil's most influential positivist engineers. During the construction of the telegraph on the eve of the Brazilian Republic, Rondon

> quickly moved beyond a purely strategic rationale for telegraph construction. For him, the key was to develop the region, to populate it with small farmers, and to build thriving towns where none currently existed. He noted of telegraph construction that 'more than the military defense of the Nation that every government seeks to secure…we have come to promote the principal necessities of populating and civilizing our Brazil' (Diacon, T., 2004, p. 132).

Primarily motivated by positivism, engineers like Rondon tried to achieve economic and political development of their new countries by significantly reorganizing and integrating indigenous and rural communities into national wholes without much (if any) concern for preserving ecosystems or local cultures. These were not concerns of the times, yet they help us understand the emphasis of engineers in constructing their national societies.

Critical Questions

As you envision your participation in a community development or humanitarian engineering project or initiative, check your assumptions about the partnering community. Do you think that they need to be better organized or connected through infrastructure (a road, a water distribution or sewage system, a computer network) to a larger whole (a village, a county, a country, a market)? If so, what might the people that you are trying to help be winning and losing through these connections?

2.3   ENGINEERS AND INTERNATIONAL DEVELOPMENT (20TH CENTURY)

During the first half of the 20th century, many engineers participated, directly or indirectly, in the building and expansion of their nation-states. In the US, for example, engineers predominantly worked in what would become the big corporations of American capitalism, such as Ford, General Motors, General Electric, DuPont, and federal and state government agencies such as the US Corps of Engineers or the Tennessee Valley Authority (TVA) (See Figure 2.3) (Hughes, T., 1989; Reynolds, T., 1991). In the USSR, engineers worked in the construction of mega-projects, like the "steel city" of Magnitosgork and the White River Dam, which came to symbolize the strength of Soviet socialism (Graham, L., 1993). In those countries that were still colonies (most of Africa and South-eastern Asia), engineers still worked on building and maintaining infrastructures for the benefit of empires (Adas, M., 2006). In either case, national and imperial development took precedence over local communities and the environment.

Exercise 1 *Find out who the main employers of engineers are at your school. How many of those corporations (like GM) or organizations (like the US military) were around in the US in the early 20th century? When were the newer corporations created? What does this relationship between corporations and engineering employment tell you about engineers?*

After WWII, a new area for engineering involvement emerged on the world stage: *international development.* With a new wave of independent countries emerging in Africa and Asia, engineers engaged enthusiastically in both national and *inter*national development. Despite their political differences, engineers from the US and USSR were both motivated by the *ideology of modernization.* That is, after 1945, many American and Soviet engineers came to believe that it was possible to develop and modernize the world through science and technology, i.e., to move "traditional" societies from their current stage of backwardness and launch them through a stage of "take-off" by implementing large development projects (hydroelectric dams, steel mills, urbanization). As discussed in the Introduction, many engineers have held to this belief to this day. Political elites and technocrats in many of these "developing" countries hoped that their countries could join the superpowers in a "modern" stage of consumer capitalism (US) or industrialized socialism (USSR) (Adas, M., 2006). Quickly, this vision was institutionalized in a number of ways such as:

- **Specific postwar plans:** e.g., the Marshall Plan in Europe and the Alliance for Progress in Latin America.

- **Technical assistance agencies**: e.g., the US Agency for International Development (USAID).

Figure 2.3: TVA under construction. "Everyone who lived near the [Tennessee] river was affected by this. Tens of thousands of jobs were created. Some of the "workers' villages" that were built during dam construction still remain – Norris, Tennessee, being the best example. Thousands of homes, hundreds of farms, and many towns were permanently flooded and had to be moved to higher ground. (Source: http://www.tnhistoryforkids.org/students/h_7. Credit: TVA).

- **"Independent" regional or international development organizations**: e.g., the World Bank, the Inter-American Development Bank, and other development banks.

- **Mega development projects**: e.g., the Aswan Dam in Egypt, the Green Revolution in Southeastern Asia, and the Itaipu Dam in Brazil.

This vision was also carefully conceptualized and disseminated by economists who heavily influenced engineers' thinking, such as W.W. Rostow at MIT, and adopted by technocrats in the US, USSR, and China alike (Adas, M., 2006, ch. 5).

Key Terms

Modernization: Modernization can mean many things, depending on the context. When we think about modernization in the development context, however, we are usually talking about the belief that communities, societies, or countries can be moved, step by step, from various states of "backwardness" or "lack" to stages of increasing wealth, "civilization," and access to technology and information. The concept is open to critique because it implies, sometimes wrongly, that certain ways of living are inferior (typically the South countries) to other ways (typically Northern countries), which are seen as superior. Modernization has also been the justification for many development programs, which in some cases have left "backward" societies worse off than before they encountered "civilization."

Technocracy: Technocracy is frequently defined as a form of government that is planned, organized and run by a group of highly educated experts. Technocrats, who are often scientists, engineers and economists, approach social problems the way they approach scientific problems, by breaking them down into constituent parts and integrating technology as means of management and/or control of those parts. Technocracies are frequently criticized as anti-democratic because technocrats centralize, rather than share, the processes and knowledge needed to rule. In effect, they can make it very difficult for the average citizen to be involved in governance because the systems they devise are so complex. The 2008 collapse of the global financial system is often attributed to technocrats.

Exercise 2 *Google "USAID" and "engineers" for images. What kind of images do you get? What do you see in them? What kind of cartoons? What do these images tell you about engineers' involvement in international development?*

As depicted in Figure 2.4, the ideology of modernization views human societies as having an evolutionary pattern, which progresses from traditional to modern. Societies would be able to achieve higher stages of development by changing their economic and political systems of production and participation. According to the ideology of modernization, as societies produce and consume more, the more modern they become. Traditional ways, often found in communal life, only get in the way of "efficient" economic production and mass consumption. Local communities have to be convinced, transformed, or coerced to join the modernization path for "take-off" by abandoning their subsistence economies and increasing their extraction of natural resources and manufacturing capacity to eventually reach a stage of high-mass consumption.

At the same time, technocrats, including many engineers, viewed nature as a "national resource" to be exploited in the name of modernization. Nature was to be organized, planned, and often re-distributed efficiently to help countries move from lower to higher stages of modernization. Once again, under this ideology, engineers, communities, and nature came together in problematic ways. Whether as technocrats working on planning departments or as builders of infrastructure, engineers, directly or indirectly, tried to change communities' traditional ways and to control nature so their countries could progress on the path to modernization and development.

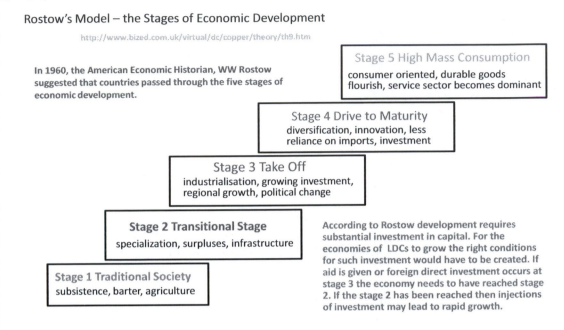

Figure 2.4: Rostow's model of economic growth. (Source: `http://welkerswikinomics.com/blog/wp-content/uploads/2008/02/growthmodels_3.jpeg` Credit: Jason Welker).

For example, during the 1960s, labeled by the United Nations as the "first development decade," engineers served in international development projects as major components of the Cold War. For example, as the US and USSR battled for influence in Egypt, US engineers built a fertilizer plant in Suez (Mitchell, T., 1988) while USSR engineers worked in the construction of the Aswan High Dam (See Figure 2.5) (Moore, C., 1994; Lotfy, et al., 2006). In Brazil, US, Italian and Brazilian engineers joined forces to build the Itaipu hydroelectric dam in 1971, one of the flagship projects of a military regime committed to contain the spread communism in Latin America. These modernization projects irreversibly changed local ecosystems and communities and enhanced the local governments' capacity to impose an ideological position on either side of the Cold War.

In spite of the powerful calls to protect nature and control human population that emerged in the 1960s (e.g., Carson's *Silent Spring*, 1962; Erlhich's *Population Bomb*, 1968), international development projects moved forward. For example, while US engineers worked on the expansion of the Green Revolution in South East Asia (Adas, M., 2006), USSR engineers participated in the "sovietization" of industrial development in the new East Germany (Stokes, R., 2000). In the case of US-financed projects, engineers' main concern was to forge a path of development towards modernization and to contain the expansion of communism, or in the case of USSR- or Chinese-financed projects, to modernize and contain the expansion of capitalism. These concerns dictated the

Figure 2.5: Built by Egyptian and Soviet engineers during the Nasser Era (1952-70), the Aswan High Dam is a clear example of engineering for development during the Cold War.
(Source: http://www.pbs.org/wgbh/buildingbig/wonder/structure/aswam2_dam.html
Credit: UPI/Corbis/Bettman. Permission Pending).

location, size, and reach of projects and neglected any consideration for environmental sustainability or autonomy of local communities (Adas, M., 2006).

Key Terms

Green Revolution: Beginning in 1945 in Mexico and then expanding to other highly populated countries like India, this revolution refers to the transformation of agriculture by means of high-yield crops brought by artificial fertilizers, pesticides, and intensive irrigation. The outcomes of this "revolution" have been highly contested, with some arguing that the technologies developed during this time have drastically improved food quality and supplies to parts of the word that need them, and others arguing that some of those approaches and technologies (e.g., use of chemicals, genetic modification, centralization of food cultivation) have been damaging to local communities, ways of life, and ecologies.

Humanitarianism: Humanitarianism is a broad term encompassing many meanings. In the context of international development, we can think of it as "systematized help," in which individuals or groups, financed by donor nations and assisted by NGOs, attempt to alleviate human suffering in the face of natural and human-caused disasters or armed conflict. Humanitarians, whether individuals or organizations, can be driven by any number of concerns—religious, ethical, social, economic, opportunistic—but typically see their mission as one of compassion and altruism while the nations that finance their efforts see their mission as part of foreign policy.

Ironically, by the late 1960s and early 1970s, engineers working within the Cold War's military-industrial complex began to express concerns for how technologies fit in local contexts. In the US, for example, a small group of engineers working at the General Electric plant in Schenectady, New York, and teaching at Rensselaer Polytechnic Institute created a group called Volunteers in Technical Assistance (VITA). They focused on the development of technologies that were simple and inexpensive to build, operate, and maintain so they could be deployed in poor villages around the world (Williamson, B., 2007). Instead of delivering large aid packages or building monumental infrastructural projects, VITA engineers believed that the key to technology transfer was in the diffusion of technical information to help villagers develop technical expertise (Darrow, et al., 1986; Pursell, C., 2003). As shown in Figure 2.6, the connection between volunteerism and the use of appropriate technologies is alive and well today, institutionalized, for example, in the program Volunteers for Prosperity, supported by USAID.

Similar approaches to enhance the technical capacity of communities were implemented in humanitarian crises by engineers like Fred Cuny, who were concerned with the welfare of people in poor regions of the world (See Figure 2.7). These people, often because of their poverty, became the most vulnerable to armed conflict, natural disasters or human-induced environmental catastrophes, famines or other grave threats to human security (Cuny, F., 1983; Cuny and Hill, 1999). A civil engineer from Texas A&M turned disaster-relief specialist, Cuny proposed a new approach in dealing with communities, as he viewed them not as passive victims of international aid but as integral partners in reconstruction efforts:

> The term victim has many negative connotations. It provokes images of helplessness, of people who must be taken care of. For this reason, many [development] agencies have used substitutes such as beneficiaries or recipients…Rather than create a new word, [I] have chosen to go with *victims*. Victims, however, are not helpless. They are capable of making intelligent choices and when special allowances are made so that victims can cope with personal losses, they can participate effectively in all post-disaster activities…the term *victim* should be coterminous with *participant* (italics in original) (Cuny, F., 1983, p. 7).

Figure 2.6: Volunteers for Prosperity Website, supported by USAID. (Source: `http://www.volunteersforprosperity.gov/` Credits: Volunteers for Prosperity).

Despite this exceptional invitation to rethink of victims of disasters as *participants*, the relationship between engineers and communities during these efforts is still one of superior to inferior (expert to non-expert or expert to apprentice) where knowledge flows mainly in one direction, from the experts. The capacities and motivations that communities have in recovering from disaster often go untapped. Also, during humanitarian crises, where time is critical in saving human lives, not much attention is paid to long-term sustainability of systems or infrastructure. Ecological concerns play second fiddle to saving human lives. Community values and short and long-term desires are also often secondary to expediency and the urgency of the moment in disaster relief crises.

In short, although humanitarian or disaster-relief engineering of this sort seems a welcome and far cry from the sorts of engineering we saw under national and international development, it still represents a sometimes problematic engineering mindset of how individuals and communities organize themselves and "work." Our point here is not to be critical of such mindsets in an anachronistic way—engineers emerged from their own social contexts and often act from within the constraints and mindsets of those contexts. Rather, our point is that it's important to be self-reflective and aware of those mindsets, so that we might also acknowledge and address their deficiencies or blind spots.

As efforts like Cuny's unfolded in the US, engineering education largely ignored these marginal developments in appropriate technology transfer or humanitarian engineering. Most engineering education initiatives, including accreditation criteria for engineering programs in place since the 1960s, were aimed at making engineering more scientific. Since the rise of the Cold War and the

Frederick Cuny in Somalia, 1992

Figure 2.7: Fred Cuny in Somalia.
(Source: `www.world.std.com/~jlr/doom/cuny.htm`   Credit: Judy DeHass).

launching of Sputnik (1957) by the USSR, the dominant concern in the competencies of engineers has been mastery of the engineering sciences (Seely, B., 1999). According to a 1968 statement by the American Society of Engineering Education (ASEE), "all courses that displace engineering science should be scrutinized. The most important engineering background of the student lies in the basic sciences and engineering sciences" (American Society of Engineering Education, 1968). ABET accreditation criteria quickly and decisively came to reflect this emphasis on science. Math, basic science, and engineering science and analysis were set to take up about 80% of the engineering curriculum, with design and humanities/social sciences taking a distant second place. Thus, the decade of the 1960s in the US ended with a scientific engineering education void of any significant impetus for reaching out to "Third World" villages through technology transfer.

Post Sputnik Engineering Curriculum

Post Sputnik engineering curriculum was organized around the following main categories (in bold):

Math and basic sciences:

- Calculus, Differential Equations, Chemistry, Physics.

Engineering Sciences:

- Mechanics of solids,

- fluid mechanics,

- thermodynamics,

- transfer and rate mechanisms,

- electrical theory,

- properties of materials.

Analysis and Design
Humanities and Social Sciences
Electives

Exercise 3 *Calculate the number of credits required in your engineering major in each of the main categories of the engineering curriculum: math and basic sciences, engineering sciences, design, humanities/social sciences, electives. Calculate the percentage of the total number of credits that each category represents in your curriculum.*

- *What category is the most dominant?*

- *Which one is the least dominant?*

- *How much emphasis is there in your curriculum on courses related to community development or humanitarian engineering?*

- *In which category are these courses located?*

- *Might these courses be located in categories considered by engineering faculty and students as "soft" or "easier"?*

- *What does this exercise tell you about the relevance of engineering knowledge for community development?*

2.4   ENGINEERS AND THE QUESTIONING OF TECHNOLOGY (THE 1970s)

The 1970s began in the US with a paradox about technology. On one hand, the US demonstrated its technical superiority to the USSR with the Apollo moon landing in 1969 (See Figure 2.8). At the same time and for a variety of historical reasons, there emerged a sharp rise in the questioning of the military-industrial complex, the impact of industrial technologies on the environment, and the use of military technology in the Vietnam War. In both popular and scientific media, science and engineering were both exalted for their achievements *and* questioned for their lack of relevance to solve domestic problems (Cass, J., 1970; Heilbroner, R., 1970). Efforts at making science and engineering relevant to society pressured companies and government agencies to find ways to apply military technologies, such as the systems approach (Dyer, D., 2000), and academic research and development (R&D) to societal problems like poverty eradication and urban renewal (Gershinowitz, H., 1972).

Figure 2.8: Engineers working in the launch control center preparing for the launch of Apollo XI. (Source: `http://upload.wikimedia.org/wikipedia/commons`).

On the international stage, the United Nations and other international organizations shifted their approach to development toward fulfilling basic needs and eradicating poverty. First proposed by World Bank's president Robert McNamara in 1972, the basic needs approach was an attempt "to reconcile the 'growth imperative' with social justice by sketching a dramatic picture of the conditions of people in the South, who were unable to take their destiny into their own hands because they could not satisfy their 'most essential needs'" (Rist, G., 2004, p. 162). After almost two decades of institutionalized international development, proponents of the "basic needs" approach wanted reassurance that development assistance was actually reaching the poorest of the poor without much interference from international bureaucracies or local governments. Yet, as historian of development Gilbert Rist points out, "Even if the fundamental case for development is a moral one [as in the case of basic needs], *the ultimate goal was to raise the productivity of the poorest so that they could be brought into the economic system.*" (his italics, Ibid., p. 163). Equally troubling in this approach is how it reinforces the notion that poor people are "unable to take their destiny into their own hands."

What Rist means is that, under a "basic needs" approach, local communities—with their differences in culture, geography, demography, etc.—are also reduced to basic needs in shelter, food, water with the goal of making them productive and incorporate them into the economy. By focusing on basic needs, development technocrats, including engineers, viewed communities strictly in terms of their deficits (water, food, shelter), instead of valuing their assets, capacities, and diversity. A "basic needs" approach encourages engineers to view communities in terms of *deficiencies* and to use universal parameters (e.g., minimum body temperature; maximum number of days without water or food, etc.) as boundary conditions for their designs (See Figure 2.9). Although development historians such as Rist claim that the basic needs approach ended with the decade of the 1970s, the approach was still advocated in the late 1990s[2] and is still alive among present-day humanitarian engineers who use the approach to energize students to join in their quest to alleviate poverty. Actually, the current vision of EWB-USA calls for "a world in which all communities have the capacity to meet their basic human needs" (`http://www.ewb-usa.org/AboutUs/VisionMission/tabid/62/Default.aspx`).

Exercise 4 *Find out how your faculty and student peers involved in community development or human- itarian engineering might be thinking about the human body. In their designs, do they envision it as a machine constrained by certain physiological parameters that can be quantitatively measured? Or as a body of mass that exchanges energy with its surrounding environment? As absent altogether from designs? Or perhaps as something else? Where do the numerical values of the parameters under consideration come from? Are these ergonomic and/or physiological values obtained by averaging those of controlled groups such as US Army soldiers or participants in medical experiments in rich countries? If so, how appropriate are these assumptions and values when designing for diverse groups of people in different parts of the world? More broadly, in community development or humanitarian engineering, what are the advantages and disadvantages of a focus on human (mostly physiological) needs?*

 The questionable outcomes of the Green Revolution, and particularly the negative impact of fertilizers and monocultures on ecosystems and local economies, brought widespread attention, prob- ably for the first time, to the long-term sustainability of large-scale development projects (Pearse, A., 1980). The "social and environmental impact" and appropriateness of technology to local settings and communities also gained widespread attention thanks to books like economist E.F. Schumacher's *Small is Beautiful* (Schumacher, E., 1973).

 A few engineering societies and schools organized conferences linking appropriate technol- ogy and development (Cook, J., 1973; American Society of Civil Engineers, 1978), while some US universities created programs in appropriate technology, as was the case at the University of Cali- fornia at Davis (Pursell, C., 1979), and science, technology and society (STS) programs. Many of these programs were developed in conjunction with engineering faculty and attracted some engi- neering students who were concerned with the social and environmental impacts of technology (e.g., Stanford, Cornell, SUNY Stony Brook, Penn State, Lehigh, MIT, Virginia Tech, and Rensse- laer) (Cutcliffe, S., 1990). Yet for the most part, the questioning of technology and its appropriateness to different local settings remained outside of mainstream engineering education.

---

[2]See *An Assault on Poverty: Basic Human Needs, Science and Technology* By IDRC, United Nations. Commission on Science and Technology for Development. Panel on Technology for Basic Needs, International Development Research Centre (Canada), United Nations Conference on Trade and Development Published by IDRC, 1997.

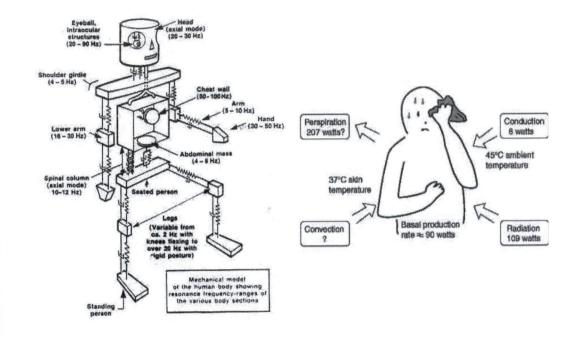

Figure 2.9: Through simplifications like these ones, engineers often depict the human body as a mechanism made of multiple components and fixed parameters such as resonance frequencies, heat transfer from different parts of the body, etc.
(Source: http://www.powerstandards.com/FunStuff/HumanResonance/HumanResonance.htm
Credit: Sven-Olof Emanuelsson.
Source: //hyperphysics.phy-astr.gsu.edu/HBASE/thermo/imgheat/bodycool3.gif).

In short, in the 1970s, appropriateness and social and environmental impact emerged as concerns for at least a few engineering professionals, educators, and students. Communities and nature became more visible here, yet communities were redefined by development technocrats in terms of *basic needs*. For engineers who had been advocating for solar energy since the 1960s, the oil embargoes and energy crises of the 1970s opened a small opportunity for engineers to get involved in the development of renewable energy and hence in an early form of sustainability. Unfortunately, this opportunity was short-lived. In the 1970s U.S., most engineers worked in companies that depended heavily on the production and/or consumption of fossil fuels and other petroleum-based products (e.g., auto-manufacturers, GE, Boeing, DuPont). The election of Ronald Reagan to the US presidency closed any possibility of federal funding for renewable energy or appropriate technology transfer to the "Third World" (Laird, F., 2001, Friedman, T., 2008, p. 14). In the US, as we will see, the institutional and political contexts of the 1980s worked against any significant development in the relationship among engineering, communities, and what would later be called sustainability.

2.5 ENGINEERS AND THE "LOST DECADE OF DEVELOPMENT" (THE 1980s)

In the 1980s, the rise of neoliberal economics and the decline of the Cold War altered the course of international development. Neoliberal policymakers placed their faith in free markets and the individual decisions of producers and consumers, arguing for a reduction of government regulations in the marketplace and the privatization of many public services. These policymakers argued that the market, not the state, ought to decide what is best for education, health, technological innovation, and international development. In the US, the election of President Reagan sparked the elimination of governmental programs for appropriate technology, such as Appropriate Technology International (ATI), part of USAID, and science and engineering programs for societal needs (Lucena, J., 1989). The AT movement suffered the consequences of this political shift (Winner, L., 1986, Ch. 6).

Key Terms

Neoliberal economics: An economic ideology that 1) endorses the free-market as the ultimate authority of who wins and who loses in the economy, 2) calls for the privatization of public services so these become part of the free market, and 3) advocates against any regulation by the government on the economy. Since the rise of neoliberal economics in the 1980s, many in the North and South have come to question the assumption that the market is a "neutral" or even "rational" arbiter of economic relations. See Saad-Filho and Johnston (2004); Greenhouse, C. (2009); Martinez, M. (2009).

Structural adjustment: The development policy of neoliberal economics where development banks and lending institutions (e.g., World Bank and IMF) make privatization, deregulation and reduction of trade barriers as conditions for "developing countries" getting new loans or reduced rates on existing loans.

The rise of neoliberal economics in many parts of the world brought a transformation of international development by eliminating the basic-needs strategy and forcing countries into policies of *structural adjustment*, where most social programs in health, education and employment would be significantly reduced, eliminated or transferred to the private sector. International development programs focused on poor national governance, reducing government intervention, shifting control of public services from the state to the private sector, and hence increasing privatization. Local communities often became disempowered as they faced the challenges of free-markets under unequal competition and the diminishing of state functions, mainly health, education and other forms of social protection. Environmental regulations came under attack as examples of government intervention on what should otherwise be a place for the free market to decide what is best: the use of natural resources (Rist, G., 2004, Ch. 10).

Consequently, the UN labeled the 1980s as the "lost decade for development" after "employment and basic needs strategies…incorporated in the Third Development Decade Strategy [the 1970s] were swept off the global and national agendas" (United Nations Intellectual History Project, 2005).

Exercise 5 *Visit the projects website for the World Bank. Under "Advanced Search," type "structural adjustment" as keywords or phrase. Likely, you will get more than 1000+ projects. Browse through the list. What do you see? What are the projects about? Where are they located? Read in detail a couple of projects that interest you. Take note of the goals of the project and how even infrastructure projects might be trying to dictate local economies, governance, private vs. public sector balance.* http://web.worldbank.org/WBSITE/EXTERNAL/PROJECTS/0,,menuPK: 115635~pagePK:64020917~piPK:64021009~theSitePK:40941,00.html

In this new political and ideological environment, engineering and engineers rose to center stage. As the US government and businesses began defining new national challenges in terms of economic competitiveness against rising technological threats such as Japan and Korea, engineers emerged as the new warriors that would help the US beat the Asian "dragons" and "tigers" in the technological marketplace. Although important discussions were taking place on the tension between economic growth and the environment, most importantly those that lead to the Brundtland Report (produced by the UN-appointed World Commission on Environment and Development in 1987), US engineering education and practice remained detached from that debate. Instead, engineering education focused on manufacturing, CAD/CAM, and the recruitment of more and more engineers to beat emerging Asian economies in the global economy (MIT Commission on Industrial Productivity, 1989; Downey, G., 1998). With the disintegration of the USSR and the end of the Cold War, other countries joined the bandwagon of economic competitiveness, including the former communist countries of Eastern Europe, which focused on reconstruction of their Soviet-age infrastructures and economies to "catch up" with the West (Hart, J., 1992; Pudlowski, Z., 1997). As engineering societies and educators became preoccupied with enhancing the economic competitiveness of their nations, the brief impetus for appropriateness and socio-environmental impact of technology achieved during the 1970s was lost to the geopolitical and ideological realities of the 1980s.

Ironically, these concerns over economic competitiveness brought the rise of engineering design education in the early 1990s. Design courses were first legitimized as countering overly theoretical engineering curricula that produced inflexible engineers incapable of competing in a global marketplace (Lucena, J., 2003). The first concerted push to incorporate flexibility in engineering education and to graduate flexible engineers came in 1990 from an NSF/NAE-sponsored workshop entitled "Engineering, Engineers, and Engineering Education in the 21st century." Engineer Roland Schmitt, at the time President of Rensselaer, chairman of the National Science Board, and the workshop's chairman, questioned the emphasis on engineering sciences in place since the 1960s:

"the unanticipated consequences of emphasis on engineering science were to ignore manufacturing, to focus on sophistication of design and features, and less on cost and quality. Some of the engineering education decisions made in the past had detrimental effects on competitiveness…We need to develop a more flexible definition of 'engineers' and 'engineering'." (Schmitt, R., 1990).

To become flexible, US engineering students needed more experience in design (Downey and Lucena, 2003). For almost two decades now, engineering design faculty committed to reforming curricula have battled for more space for design courses. As expected, the design models and practices that emerged were for industry, not for community development; hence, they contained many problematic assumptions about the ways engineers have engaged communities through their designs:

1. Design projects should strengthen connections between engineering schools and private industry (not local communities).

2. The relationship between engineers (students supervised by faculty) and those in "need" of a product parallels that of expert and clients (not as equal partners in a collaboration).

3. Budgetary and legal constraints should be considered high priorities in design considerations (instead of ecological sustainability and community empowerment).

4. Through design education, students will become "flexible" in a competitive marketplace and more ready for jobs in industry (not listeners and facilitators in community development).

5. Team-work is viewed as division of labor among students of different engineering disciplines and forms of expertise or knowledge (not as partnership with people who hold different perspectives than your own).

In Chapter 3, we discuss how most community development and humanitarian engineering initiatives that have come to rely on existing engineering design courses have inherited some of these problematic assumptions. We will expand on this tricky relationship throughout the book, particularly as it affects engineers in ESCD projects.

2.6   ENGINEERS MOVE TOWARD SUSTAINABLE DEVELOPMENT (1980s-1990s)

Sustainable development was a trend that developed largely out of the failures of the development strategies of the 1970s and 1980s. One of the key events in this history was the 1992 United Nations Conference on Environment and Development in Rio de Janeiro (also known as the Earth Summit), out of which came the Rio Declaration.

We have identified two dominant views of sustainable development—the weak and the strong (Neumayer, E., 1999). *Weak sustainability*, also called "constrained growth," emphasizes economic models that do not differentiate between natural and human-made resources. Proponents of this view assume that scientific and technological advancement will address natural resource depletion and emphasize the importance of economic and social gains in the face of environmental degradation. Due to its reliance on technological solutions, most engineers have traditionally supported this approach (See Figure 2.10).

By contrast, proponents of *strong sustainability* acknowledge that natural resources cannot always be treated like human-made resources because of natural constraints such as irreversibility of ecological damage (e.g., you cannot bring an animal species back to life once it is gone). This view argues for the protection of natural resources even at the cost of development opportunities (e.g., saving the spotted owl even if it means losing growth opportunities for the timber industry).

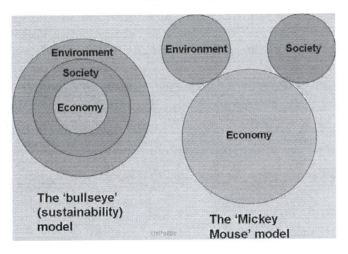

Figure 2.10: Strong sustainability can be depicted with the economy as dependent on social and economic activity which in turn are dependent of the natural environment. Activities that are harmful to the environment damage both society and the economy (the "bullseye" model). Weak sustainability can be represented with the economy as the main focus of human activity and both society and the environment as relevant but tangential considerations. In the 'Mickey Mouse' model, protecting the environment might be desirable but not essential to society or the economy.
(Source: http://www.ozpolitic.com/articles/environment-society-economy.html Credit: OzPolitic).

Key Terms

Weak sustainability: This conception of sustainability sees natural resources much the way we see economic ones—as something to be priced, bought, sold and managed. It views nature in terms of markets, economic worth, and technocratic management. Its appeal is that it does little to challenge prevailing beliefs about economic growth and human consumption, assuming that natural resources can simply be incorporated into existing economic models. Its disadvantage is that it doesn't take into account important characteristics that make natural resources different from human-produced resources such as their finite nature and our utter dependence on them for survival. The cap and trade approach to CO_2 emissions is an example of weak sustainability.

Strong sustainability: This model assumes that environmental or natural resources have intrinsic value in relation to other forms of capital and human-made resources. While pollution is often "externalized" in the weak model, it would be accounted for in the strong model because it represents damaging of natural capital, or the commons. The advantage of this model is that it makes good ecological sense; we cannot have a timber industry, for example, if there are no trees to harvest due to over-logging. On the other hand, it has proven very difficult to change the economic system to include "externalities" because the weak model is so in line with our deeply ingrained assumptions about what has worth.

Lacking the nationalistic luster of economic competitiveness, which placed engineers at the center stage of technological innovation in the 1980s, sustainable development was only a marginal preoccupation for engineers in the 1990s. Among a myriad of reports linking technological development to economic competitiveness, one on *Technology and Environment*, by the US National Academy of Engineering (NAE), called for "[engineers as] creators of new technological developments and policymakers…to develop guidelines and policies for sustainable development that reflect for the long-term, global implications of large-scale technologies and that support the innovation of less intrusive, more adaptable technologies at all levels" (Ausubel and Sladovich, 1989).

Despite such calls, sustainable development did not provide the market demand that would justify investments in new sustainable technologies. By contrast, economic competitiveness clearly challenged engineers to develop technologies for ever growing international markets. Most corporate employers of US engineers were simply not willing to take sustainable technology investment risks. New markets for sustainable technologies had to be created with government incentives and through policy decisions such as those highlighted by President Clinton's Council on Sustainable Development (1993-96) (Zwally, K., 1996). Unfortunately, neither the Clinton nor the Bush administrations provided sufficient incentives to create these markets. It remains to be seen whether the commitment of the Obama administration towards renewable energy materializes in such markets, products, and jobs—which could attract future generations of engineers.

In engineering education, sustainable development did not become a major theme in the 1990s, marginally appearing through the concerns of a small community of activist engineering educators that annually puts together the International Symposium on Technology and Society (ISTAS) of the Institute of Electrical and Electronics Engineers (IEEE). In 1991, ISTAS held a symposium entitled "Preparing for a Sustainable Society." Sustainable development became a theme around which a handful of engineering educators proposed new curricula in engineering ethics, economics and the academic field known as science, technology, and society (STS) (IEEE, 1991). Unfortunately, at that time, these proposals became secondary in engineering programs, largely because economic competitiveness was challenging most engineering faculty to focus curricular development in areas that US engineering students seemed to be lacking, such as design, manufacturing, and international education. The calls for "flexible engineers" that would help the US compete in a global economy did not include competencies related to sustainable development (Lucena, J., 2003).

2.7   ENGINEERS HEED THE CALL TO SUSTAINABLE DEVELOPMENT (LATE 1990s-PRESENT)

In contrast to the preceding decades, engineering organizations in the early 21st century heeded the call to sustainable development and have begun taking actions, ranging from hosting regional and world conferences to declaring their position with respect to sustainable development, to revising their codes of ethics and challenging members to address sustainable development principles in their work, and creating international professional partnerships such as the World Engineering Partnership for Sustainable Development (WEPSD). The WEPSD vision statement indicates that

> Engineers will translate the dreams of humanity, traditional knowledge, and the concepts of science into action through the creative application of technology to achieve sustainable development. The ethics, education, and practices of the engineering profession will shape a sustainable future for all generations. To achieve this vision, the leadership of the world engineering community will join together in an integrated partnership to actively engage with all disciplines and decision makers to provide advice, leadership, and facilitation for our shared and sustainable world (World Federation of Engineering Organisations, 1997, p. 7).

In 1999, the American Society of Engineering Education (ASEE) released a "Statement on Sustainable Development Education" which states that

> Engineering students should learn about sustainable development and sustainability in the general education component of the curriculum as they are preparing for the major design experience. For example, studies of economics and ethics are necessary to understand the need to use sustainable engineering techniques, including improved clean technologies. In teaching sustainable design, faculty should ask their students to consider the impacts of design upon U.S. society, and upon other nations and cultures. Engineering faculty should use systems approaches, including interdisciplinary teams, to teach pollution prevention techniques, life cycle analysis, industrial ecology, and other sustainable engineering concepts.... ASEE believes that engineering graduates must be prepared by their education to use sustainable engineering techniques in the practice of their profession and to take leadership roles in facilitating sustainable development in their communities" (ASEE Board of Directors, 1999).

In addition, as a part of its code of ethics, the American Society of Civil Engineers (ASCE) has declared that its engineers shall "strive to comply with the principles of sustainable development," which is defined as "the challenge of meeting human needs for natural resources, industrial products, energy, food, transportation, shelter, and effective waste management while conserving and protecting environmental quality and the natural resource base essential for future development." Other professional societies and organizations have followed suit.

Although sustainable development did not challenge engineers to compete in the international arena in the same way that economic competitiveness has done since the 1990s, it became an interesting problem for some engineers to solve through a *systems approach*. Some engineers appropriated "sustainable development" as an effort to be achieved through the use of technologies to clean up the mess that previous industrial practices had created and positioned themselves as "central players" in the success or failure of this effort (Prendergast, J., 1993). Ironically, the systems approach that emerged in the 1950s out of military technological development (Hughes, et al., 2000) was favored again as a key engineering tool to solve the challenges of sustainable development. This systems approach to sustainability has become institutionalized in a small number of engineering education programs such as University of Michigan's Engineering Sustainable Systems dual-degree (see http://www.snre.umich.edu/degree_programs/engineering). Figure 2.11 is an example of a systems approach to modeling a lake that reveals the complexity of the relationship among biophysical and socio-economic parameters.

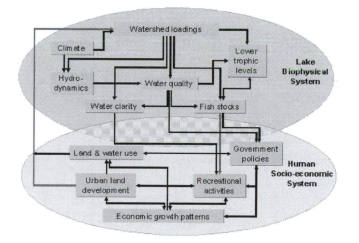

Figure 2.11: Modeling of coupled parameters in a lake system (Fiksel, J., 2006).

This is a welcome improvement in engineers' understanding of how human systems interact with ecological ones. Yet excessive analysis of these interactions can lead to inaction. In his excellent summary of systems approaches to sustainability, including those developed by engineers and other scientists, Joseph Fiksel warned us that "[w]hile improving modeling techniques and establishing a rigorous science of sustainability is important, a caveat is in order. Excessive modeling efforts may become an excuse for delaying effective political action, leading to 'paralysis by analysis'…. Progress in theory-based research needs to be balanced with exploratory policy implementation that will enrich our understanding of sustainability issues in real-world systems" (Fiksel, J., 2006, p. 20).

As the end of the 20th century approached, some engineering educators incorporated sustainable development in the desired set of knowledge and skills for the engineer of the 21st century (Velazquez, et al., 1999). The emergence of new ABET accreditation criteria for engineering programs in the US in 2000 facilitated this adoption, especially the criterion that calls for engineering graduates to have "an ability to design a system, component, or process to meet desired needs within realistic constraints such as economic, environmental, social, political, ethical, health and safety, manufacturability, and sustainability." Furthermore, the influential *Engineer of 2020* report challenges engineers in the 21st century to adopt the tools for sustainable designs to the local conditions of developing countries in order to ensure equity in the benefits from using these tools across the world (National Academy of Engineering, 2004, p. 21).

Despite these commitments to sustainable development, there is little evidence showing that most engineering students are learning about it. Although engineering students nowadays seem to show more awareness of environmental issues, they lack knowledge of definitions of and approaches to

- sustainable development,

- key sustainable development principles and concepts such as the precautionary principle and inter- and intra-generational equity,

- social justice in general,

- and how to deal with stakeholder participation in sustainable development (Azapagic, et al., 1999).

In a recent workshop on engineering design and sustainability, education researchers confirmed that students see the application of tools for sustainability, such as Life-Cycle Assessment (LCA), and the practice of engineering as contradictory:

> They [students] expressed that the particular focus on LCA would mean that 'functionality is made secondary' or that they would have to 'only think of the environment', which students expressed as a puzzle or contradiction to their understanding of engineering. The LCA is perceived as a borderline engineering related task. The researchers did not see much evidence that environmental issues are perceived as a required component of what makes a product 'functional'. A different version of the same argument surfaces, when students express LCAs are more valuable for end-users and less valuable for engineers (Strobel, et al., 2009, p. 11).

This book cannot address all of these knowledge gaps; but it hopes to provide plausible answers as to why these gaps exist. We will analyze how traditional engineering design courses might be contributing to these knowledge gaps in Chapter 3.

Exercise 6 *Following Azapagic et al's survey of engineering students (Azapagic, et al., 1999), assess your own knowledge of the following topics related to sustainable development:*

- *Intergovernmental Panel on Climate Change (IPCC)*

- *ISO 14001*

- *Kyoto Protocol*

- *Montreal Protocol on CFCs*

- *Rio Declaration*

- *Eco-labelling*

- *Industrial ecology*

- *Product Stewardship*

- *Tradable permits*

- *Precautionary principle*

- *Inter- and intra–generational equity*

- *Stakeholder participation*

Use a scale of 0 to 4 where 0= not know; 1= know a little bit; 2= know somewhat; 3= know quite a bit; 4= know a lot.

2.8   THE EXPLOSION OF "ENGINEERING TO HELP" (ETH) ACTIVITIES (2000-PRESENT)

Since the early 1990s, engineering activities dealing with humanitarian and community development activities have proliferated significantly. Stimulated by the involvement of other professions in humanitarian relief, such as Doctors Without Borders (1971), Reporters Without Borders (1985), and Lawyers Without Borders (2000), engineers took up the challenge and independently organized a number of groups under some form of the name "Engineers without Borders": France's Ingénieurs Sans Frontieres (late 1980s), Spain's Ingeniería Sin Fronteras (1991), Canada's Engineers Without Borders (2000), Belgium's Ingénieurs Assistance Internationale (2002), and others. In 2003 these groups organized "Engineers Without Borders-International" as a network to promote "humanitarian engineering ... for a better world," now constituted by more than 41 national member organizations (http://www.ewb-international.org/members.htm).

Simultaneously, many other engineering activities trying to address the challenges of sustainable development have emerged. There are now many student organizations and academic initiatives, such as those listed in the Introduction, NGO-driven organizations such as Engineers for a Sustainable World (ESW) and journals, such as *Environment, Development and Sustainability* (2002-present), *Engineering Sustainability* (2003-present), and *Journal of Engineering for Sustainable Development* (2006-present). This surge of activities is taking place at the historical convergence of three key events:

- The globalization of US engineering education (Lucena, et al., 2008),

- the transformation of long-term corporate loyalty to engineering employees (Barley and Kunda, 2004),

- and the unparalleled media coverage of humanitarian crises, violent conflict, poverty, and environmental degradation occurring worldwide (Hoijer, B., 2004).

Let us briefly analyze this historical convergence.

As we have seen, the end of the Cold War and the new challenge of global economic competitiveness brought significant changes to US engineering education, including a redefinition of engineering competencies embodied in the ABET EC 2000 criteria (Lucena, J., 2003). The new engineering competencies, intended in part to create global engineers out of US-educated engineers, "has also provided opportunities to other programs and organizations not explicitly aimed at producing competencies for industry" such as EWB, ESW, etc. (Lucena, et al., 2008, p. 5). In short, ETH initiatives have emerged at an opportune time, when engineering programs still struggle to address challenges of ABET accreditation such as developing the abilities "to design a system to meet desired needs…to function in multidisciplinary teams…to understand professional and ethical responsibility…[and] to understand the impact of engineering solutions in a global context" (ABET, 2002).

Also, since the 1980s, engineers have been experiencing significant dislocations in corporate employment. Practices aimed at increasing work productivity (i.e., more output per unit of human labor) put in place since the 1980s have resulted in continuous cycles of layoffs, workplace restructuring and geographic job reallocations from the US to countries like China and India (Aronowitz and DiFazio, 1994; Rifkin, J., 1995; Friedman, T., 2006). No longer committed to their corporate employers, increasing numbers of engineers have become "itinerant experts in a knowledge economy" outside of mainstream employment (Barley and Kunda, 2004). These dislocations of engineering employment of the last two decades have opened opportunities for many engineers to serve the public beyond the constraints set in place by many years of corporate employment by volunteering and/or even seeking employment as "relief engineers" (Davis and Lambert, 1995) in humanitarian, community development or sustainable development organizations (For an extensive analysis of this emergence, see (Schneider, et al., 2009).

Key Terms

North-South Divide (or **Rich-Poor Divide**): Proposed as a more accurate division of the world than the widely (mis)used First-Second-Third Worlds division, this socio-economic division shows the economic gap that exists between the wealthy countries known collectively as "the North," and the poorer countries, or "the South." Although most nations comprising the "North" are in fact located in the Northern Hemisphere, the divide is not only defined by geography but has come to reflect political power in the world stage. The North is home to four out of five permanent members of the UN Security Council and all members of the G8.

Also in the last few decades, we have witnessed an unprecedented increase in media portrayals and coverage of humanitarian crises around the world. Beginning with the first televised famine in Biafra (1967), those around the world with access to TV have seen the graphic images of human suffering during the conflicts in Vietnam, Kosovo, Rwanda, Kurdistan, Palestine, Chechnya, and Darfur and after disasters like the tsunami in Indonesia and hurricane Katrina, to name a few. This media exposure, coupled with enduring ideas of progress and superiority of the North over the South, have produced what Barbara Heron calls "a planetary consciousness" and "a sense of entitlement and obligation to intervene globally." She argues that this sense of entitlement and obligation explains "why middle class Americans respond to media portrayals of global problems by feeling, as [Edward] Said argues, that it is up to them to set right the wrongs of the world…" (Heron, B., 2007, p. 37). Engineers have not remained distant from this exposure and appropriation of images of the poor and dispossessed (See Figure 2.12). Often during speeches or ETH program brochures, humanitarian engineers justify their sense of entitlement and obligation to help others by summarizing the statistics of suffering (e.g., number of people without water, number of people earning 1 dollar a day… etc.) and showing pictures of the poor in the South.

Exercise 7 *What underlying assumptions regarding the North's attitude toward people in the South are (explicitly or implicitly) conveyed by the following:*

- *World Vision TV commercials (Search for these at* youtube. com*).*

- *EWB Website (See* http: // www. ewb-usa. org/*).*

- *UNICEF commercials (Search for these at* youtube. com*).*

2.9   THE EMERGENCE OF *COMMUNITY* IN SUSTAINABLE DEVELOPMENT AND ETH INITIATIVES

After many years of development failures and the emergence of sustainable development in the 1990s, some engineers and development workers, and even bureaucrats, have begun to recognize the need to engage communities in more inclusive and participatory ways. As we have seen, since

Figure 2.12: The banner of the humanitarian engineering program's website at the Colorado School of Mines shows an image that perhaps needs no explanation since, in the US, we have been socialized by the media to immediately assign meaning to a picture like this. What does this image tell you about the person standing against the wall?
(Source: `http://humanitarian.mines.edu/` Credits: Colorado School of Mines).

the relationship between engineering and development began to take shape in the 19th century, engineering work with local communities has been problematic at best. Throughout most of this history, engineers have been guided primarily by commitments to top-down planning, design, development, and implementation of projects done without consultation with communities. This attitude toward local and indigenous communities has been perpetuated and reinforced first by colonialism, then by the ideologies of positivism and modernization, and most recently by the desire to help (Escobar, A., 1995; Heron, B., 2007). Recognizing this problem, social scientists and development practitioners have been advocating participatory practices since the 1980s to include and engage communities in meaningful participation and equal partnership instead of passive receptivity of development (Salmen, L., 1987). Some have gone as far as to claim that sustainable development is unattainable without the participation and empowerment of local communities (Blewitt, J., 2008). We explore this relationship further in Chapter 4.

Yet participatory approaches to community development remain elusive to most engineering projects for a number of reasons. Historically, we have seen how engineering practices for development have emerged in alliance with specific foreign policies, located within national and international agencies and organizations, and inspired by the ideologies of positivism, modernization, and neoliberalism. We have come to realize that this history continues to shape many of the practices of engineers in development projects and the approaches that even students take toward communities.

One engineering professor involved in the development of the EWB handbook confirmed this realization when describing the language in the first edition as condescending toward communities, communicating the idea that "we will go and we will teach them [the villagers] how to be sustainable." An article on community service planning for engineering students, published in a journal of a major engineering society and written by a student leader, outlined the steps that students need to take to identify project objectives, select projects, and solicit projects. Student satisfaction and the application of engineering knowledge are paramount criteria while community participation is marginal at best (Evans and Evans, 2001). As we will see in Chapter 3, the project that received the student humanitarian top prize from a major engineering society in 2009 finally included community input at the pilot stage—*after* students in the classroom had framed the problem, decided on the design, and built a prototype.

The relatively few US engineering educators who are involved in educational opportunities in community development, humanitarian engineering and/or sustainable development have been primarily motivated by the needs of students and curricula. For example, many of these educators who want to provide students with an international experience in a "real life" situation have to comply with ABET accreditation criteria for their engineering programs, particularly those that are difficult to incorporate in engineering courses (e.g., "the broad education necessary to understand the impact of engineering solutions in a global, economic, environmental, and societal context"). In sum, engineering educators and administrators might be supporting ETH programs and initiatives in order to

- increase student recruitment and retention, particularly of women who seem to demand more that engineering be relevant to societal problems,

- comply with accreditation criteria,

- enhance students' international and team-work experiences,

- and address increasing focus on engineering ethics (Manion, M., 2002).

These are worthy and noble causes, but they potentially place the participatory role of communities as secondary. As one committed engineering professor with many years of experience in student-led community projects recently confided to us,

> What I found is people in the villages are smart, they know what's happening, they know what they need. They may not have the funds to do certain things that they want to do, but you know this whole thing of going and doing—all this is actually benefiting our students more [than the villagers] because it's opening [the students'] eyes. So let's be honest and say 'Yeah it's a good international exposure for our students but do you want to risk these communities?' I don't know. I don't know. I seriously don't know....I still wonder if [we] left [the villagers] alone, if they would be fine.

Sustainable development and ETH programs that do not shine a critical, self-reflective light on their work may risk replicating the dangers found in this historical relationship between engineers and development which, for the most part, has disempowered the communities that engineers were meant to serve. We hope that this book will provide guidance on how to be critical and self-reflective when trying to bring engineering knowledge and skills to the service of community. Via the case studies, we also hope the book shows how engineers can listen to and engage communities in effective ways.

2.10   SUMMARY[3]

Historical period	Engineers' primary emphasis	Engineers' main view of community
Engineers and the development of empires (18th and 19th centuries).	To transform nature into a predictable and lasting machine that could be controlled to ensure their imperial patrons a return on investment and display superiority over indigenous people.	Communities as sources of potential imperial subjects to be organized in ways that made it possible to tax them, convert them to the religion of the empire and often force them into labor for the construction of imperial projects.
Engineers and national development (19th to 20th centuries).	To map territory and natural resources of new countries; to build national infrastructures to connect dispersed populations into a national whole and integrate their productive capacity for national and international markets.	Communities as part of a larger national whole (national subjects) that needed to be brought into functional *order* with other parts of the nation to ensure its *progress*.

[3]These are broad historical generalizations that perhaps apply more to engineers from certain countries than from others. For example, beginning in 1980s concerns about economic competitiveness with Japan were more prevalent among US engineers than among engineers from other countries. See Lucena, J. (2005). Defending the Nation: US Policymaking in Science and Engineering Education from Sputnik to the War Against Terrorism. Landham, MD, University Press of America.

Historical period	Engineers' primary emphasis	Engineers' main view of community
Engineers and international development (20th century).	To develop and modernize the world through science and technology; to move "traditional" societies from their current stage of backwardness and launch them through a stage of "take-off" by implementing large development projects (hydroelectric dams, steel mills, urbanization).	Communities as obstacles to "efficient" economic production and mass consumption. Local communities to be convinced, transformed or coerced to join the modernization path by abandoning their subsistence economies, increasing their extraction of natural resources and manufacturing capacity to eventually reach a stage of high-mass consumption.
Engineers and the questioning of technology (the 1970s).	Development engineers focused on providing communities' *basic needs* in shelter, food, and water with the goal of making them productive and incorporating them into the economy.	Communities viewed in terms of what they lacked (*deficiencies*) and humans in terms of basic need parameters (e.g., minimum body temperature; maximum number of days without water or food, etc.).

Historical period	Engineers' primary emphasis	Engineers' main view of community
Engineers and the "lost decade of development" (the 1980s).	Most US engineers began to embrace economic competitiveness as Japan emerged as a technological threat; development engineers engaged in structural adjustment, i.e., expansion of free markets, reduction of government regulations in the marketplace, and encouraging privatization of public services.	Local communities disempowered as they faced the challenges of free-markets under unequal competition and the diminishing of state functions, mainly health, education and other forms of social protection.
Engineers move toward sustainable development (1980s-1990s).	Most in US continued to embrace economic competitiveness; few began to consider sustainable development through a systems approach but mainly in its "weak" form.	Same as in the 1970s and 1980s.
The explosion of "Engineering to Help" (ETH) activities (2000-present).	Most still embrace economic competitiveness; some committed to help the poor and disposed in problematic ways.	Same as in the 1970s and 1980s but with some attempts as incorporating communities through participatory practices.

REFERENCES

ABET (2002). *Criteria for Accrediting Engineering Programs–Effective for Evaluations During the 2003-2004 Accreditation Cycle.* Baltimore, ABET. 40

Adas, M. (2006). *Dominance by Design: Technological Imperatives and America's Civilizing Mission.* Cambridge, Harvard University Press. 19, 20, 22, 23

American Society of Civil Engineers (1978). Appropriate technology in water supply and waste disposal : a workshop at the Annual Convention. Chicago, ASCE. 29

American Society of Engineering Education (1968). ASEE Goals Report. Washington, DC, ASEE. 26

Aronowitz, S. and W. DiFazio (1994). *The jobless future: sci-tech and the dogma of work*. Minneapolis, University of Minnesota Press. 40

ASEE Board of Directors. (1999). "ASEE Statement on Sustainable Development Education." http://www.asee.org/about/Sustainable_Development.cfm. 36

Ausubel, J. H. and H. Sladovich (1989). Technology and Environment. Washington, D.C., NAE. 35

Azapagic, A., S. Perdan, and D. Shallcross (1999). "How much do engineering students know about sustainable development? The findings of an international survey and possible implications for the engineering curriculum." *European journal of engineering education* **30**(1): 1–19. 38, 39

Barley, S. R. and G. Kunda (2004). *Gurus, hired guns, and warm bodies: itinerant experts in a knowledge economy*. Princeton, N.J., Princeton University Press. 40

Bazant, M. (1984). "La Enseñanza y la Practica de la Ingenieria Durante el Porfiriato." *Historia Mexicana* **33**(4): 254–297. 18

Bazant, M. (2002). *Historia de la educacion durante el porfiriato*. Mexico City, El Colegio de Mexico. 18

Blewitt, J. (2008). *Community, empowerment and sustainable development*. Totnes, Green Books. 42

Cass, J. (1970). In the Service of Man. *Saturday Review*. 28

Cook, J. (1973). *Appropriate technology in economic development*, seminar held in the University of Edinburgh under the joint auspices of the Centre of African Studies and the School of Engineering Science (Electrical Engineering), University of Edinburgh, Centre of African Studies and School of Engineering Science. 29

Cuddy, B. and T. Mansell (1994). "Engineers for India: The Royal Indian Engineering College at Copper's Hill." *History of Education* **23**(1): 107–123. 14

Cuny, F. (1983). *Disasters and Development*. New York and Oxford, Oxford University Press. 24

Cuny, F. C. and R. B. Hill (1999). *Famine, conflict and response: a basic guide*. West Hartford, Kumarian Press. 24

Cutcliffe, S. H. (1990). "The STS curriculum: what have we learned in twenty years?" *Science, Technology, & Human Values* **15**(3): 360–372. 29

da Silva Telles, P. C. (1993). *Historia da Engenharia No Brasil, Seculo XX*. Sao Paulo, Ed. Brochura. 17

Darrow, K. and S., Mike (1986). *Appropriate Technology Source Book: A Guide to Practical Books for Village and Small Community Technology*. Stanford, A Volunteers in Asia Publication. 24

Davis, J. and R. Lambert (1995). *Engineering in emergencies: practical guide for relief workers*. [S.l.], Intermediate Technology. 40

Diacon, T. A. (2004). *Stringing Together a Nation: Candido Mariano da Silva Rondon and the Construction of a Modern Brazil, 1906-1930*. Durham and London, Duke University Press. 16, 18

Downey, G. and J. C. Lucena (2003). "When Students Resist: Ethnography of a Senior Design Experience in Engineering Education." *International Journal of Engineering Education* **19**(1): 168–176. 33

Downey, G. L. (1998). *The Machine in Me: An Anthropologists Sits Among Computer Engineers*. New York, Routledge. 32

Downey, G. L. and J. C. Lucena (2004). "Knowledge and Professional Identity in Engineering: Code-Switching and the Metrics of Progress." *History and Technology* **20**(4): 393–420. 13

Duggan, L. (2004). *The twilight of equality? Neoliberalism, cultural politics, and the attack on democracy*. New York, Beacon Press.

Dyer, D. (2000). The limits of technology transfer: civil systems at TRW, 1965-1975. *Systems, experts, and Computers: the systems approach in management and engineering, World War II and after*. A.C. Hughes and T. P. Hughes. Cambridge, MIT Press. 28

Eakin, M. C. (2002). *Tropical Capitalism: the Industrialization of Belo Horizonte, Brazil*. New York, Palgrave. 14

Escobar, A. (1995). *Encountering Development: The Making and Unmaking of the Third World*. Princeton, Princeton University Press. 42

Evans, M. D. and D. M. Evans (2001). "Community service project planning for ASCE student chapters/clubs." *Journal of Professional Issues in Engineering Education and Practice* **127**(4): 175–183. 43

Fiksel, J. (2006). "Sustainability and resilience: toward a systems approach," *Sustainability: Science, Practice and Policy* **2**(2). (open access journal at `http://ejournal.nbii.org/archives/vol2iss2/0608--028.fiksel.pdf`.) 37, 38

Friedman, T. (2008). *Hot, Flat, and Crowded: Why We Need a Green Revolution*. New York, Farrar, Straus and Giroux. 30

Friedman, T. L. (2006). *The World is Flat: A Brief History of the 21st Century*, Farrar, Straus, and Giroux. 40

Galvez, A. (1996). Ingenieros e Ingenieria en el Siglo XIX. *La ingenieria civil Mexicana: Un encuentro con la historia*. C. Martin. Mexico, D.F., Colegio de Ingenieros Civiles de Mexico. 14

Gershinowitz, H. (1972). Applied Research for the Public Good-A Suggestion. *Science*. **76**:380–86. 28

Graham, L. R. (1993). *The Ghost of the Executed Engineer: Technology and the Fall of the Soviet Union*. Cambridge, Massachusetts 19

London, England, Harvard University Press.

Grayson, L. P. (1993). *The Making of an Engineer: An Illustrated History of Engineering Education in the United States and Britain*. New York, John Wiley and Sons. 16

Greenhouse, C. J. (2009). *Ethnographies of neoliberalism*. Philadelphia, University of Pennsylvania Press. 31

Haber, S. H. (1989). *Industry and Underdevelopment: The Industrialization of Mexico, 1890-1940*. Stanford, California, Stanford University Press. 17

Hart, J. A. (1992). *Rival Capitalists: International Competitiveness in the United States, Japan and Western Europe*. Ithaca, Cornell University Press. 32

Headrick, D. R. (1981). *The Tools of Empire: Technology and European Imperialism in the Nineteenth Century*. New York, Oxford University Press. 14

Headrick, D. R. (1988). *The Tentacles of Progress: Technology Transfer in the Age of Imperialism, 1850-1940*. New York, Oxford University Press. 14

Heilbroner, R. L. (1970). Priorities for the Seventies. *Saturday Review*. **53**:17–19. 28

Henderson, J. D. (2001). *Modernization in Colombia: The Laureano Gomez Years, 1889-1965*. Gainsville, FL, University Press Of Florida. 18

Heron, B. (2007). *Desire for development: whiteness, gender, and the helping imperative*. Waterloo, Ontario, Wilfrid Laurier University Press. 41, 42

Hoijer, B. (2004). "The discourse of global compassion: the audience and media reporting of human suffering." *Media, Culture & Society* **26**(4):513–31. 40

Horna, H. (1992). *Transport Modernization and Entrepreneurship in Nineteenth Century Colombia: Cisnerso & Friends*. Stockholm, Almqvist & Wiksell International. 17

Hughes, A. C. and T. P. Hughes, Eds. (2000). *Systems, experts, and computers: the systems approach in management and engineering, World War II and after.* Dibner Institute Studies in the History of Science and Technology. Cambridge, MIT Press. 37

Hughes, T. H. (1989). *The American Genesis: A Century of Invention and Technological Enthusiasm 1870-1970.* New York, Viking Press. 19

IEEE (1991). Preparing for a Sustainable Society. Proceedings of the 1991 International Symposium on Technology and Society. 35

Laird, F. (2001). *Solar energy, technology policy, and institutional values.* Cambridge; New York, Cambridge University Press. 30

Lotfy, O. E., J. Lucena, and G. Downey (2006). "Engineering and Engineering Education in Egypt." *IEEE Technology and Society* **25**(2):17–24. 22

Lucena, J. (1989). Appropriate technology: a contemporary review. Troy, Renssalaer Polytechnic Institute. 31

Lucena, J. (2003). "Flexible Engineers: History, challenges, and opportunities for engineering education." *Bulletin of Science, Technology, and Society* **23**(6): 419–435. 32, 36, 40

Lucena, J. (2009). Crear y Servir la Patria: Engineers and national progress in Mexico from Independence to post World War II. *Jogos de Identidade: os Engenheiros entre a Ação e a Formação.* M.P. Diogo. Lisboa: Colibri. 13, 14, 16

Lucena, J., G. Downey, B. Jesiek, and S. Elber (2008). "Competencies Beyond Countries:The Re-Organization of Engineering Education in the United States, Europe, and Latin America." *Journal of Engineering Education,* 1–15. 40

Lucena, J. C. (2005). *Defending the Nation: US Policymaking in Science and Engineering Education from Sputnik to the War Against Terrorism.* Landham, MD, University Press of America. 44

Lucena, J. C. (2009). "Imagining nation, envisioning progress: emperor, agricultural elites, and imperial ministers in search of engineers in 19th century Brazil." *Engineering Studies* **1**(3): 24–50. 13, 14

Manion, M. (2002). "Ethics, engineering, and sustainable development." *IEEE Technology and Society Magazine* **21**(3): 39–48. 43

Martinez, M. (2009). *The Myth of the Free Market: The Role of the State in a Capitalist Economy,* West Hartford, CT, Kumarian Press. 31

MIT Commission on Industrial Productivity (1989). *Made in America: Regaining the Productive Edge.* New York, Harper. 32

Mitchell, T. (1988). *Colonising Egypt*. Cambridge, Cambridge University Press. 14, 22

Moore, C. (1994). *Images of Development: Egyptian engineers in search of industry*. Cairo, The American University of Cairo Press. 14, 22

Mrazek, R. (2002). *Engineers of Happy Land: Technology and Nationalism in a Colony*. Princeton, Princeton University Press. 13

Nachman, R. G. (1977). "Positivism, Modernization, and the Middle Class in Brazil." *Hispanic American Historical Review* **57**(1): 1–23. 17

National Academy of Engineering (2004). *The Engineer of 2020: Visions of Engineering in the New Century*. Washington, DC, The National Academies Press. 38

Neumayer, E. (1999). *Weak versus strong sustainability exploring the limits of two opposing paradigms*. Cheltenham, UK; Northampton, MA, USA, E. Elgar. 33

Pearse, A. (1980). *Seeds of plenty, seeds of want: social and economic implications of the Green Revolution*. Oxford, Clarendon Press. 29

Prendergast, J. (1993). "Engineering sustainable development." *Civil Engineering, ASCE* **63**(10):39–42. 37

Pudlowski, Z. (1997). "3rd East-West Congress on Engineering Education-Re-vitalizing Academia/Industry Links." *Global Journal of Engineering Education* **1**(1):1–12. 32

Pursell, C. (1979). "The history of technology as a source of appropriate technology." *The Public Historian* **1**(2):15–22. 29

Pursell, C. (2003). *Appropriate technology, modernity and U.S. foreign aid*. International Congress on the History of Sciences, Mexico City, Mexico. 24

Regnier, P. and A.F. Abdelnour (1989). *Les Saint - Simonies en Egypte*. Cairo, Arab World Printing House. 14

Reynolds, T., Ed. (1991). *The Engineer in America: A Historical Anthology from Technology and Culture*. Chicago, University of Chicago Press. 19

Rifkin, J. (1995). *The end of work : the decline of the global labor force and the dawn of the post–market era*. New York, G.P. Putnam's Sons. 40

Rist, G. (2004). *The History of Development from Western Origins to Global Faith*. London, Zed Books. 28, 32

Saad-Filho, A. and D. Johnston *Neoliberalism: A Critical Reader*. London, Pluto Press. 31

Safford, F. (1976). *The Ideal of the Practical: Colombia's Struggle to form a Technical Elite*. Austin and London. University of Texas Press. 16

Salmen, L. F. (1987). *Listen to people: Participant-observer evaluation of development projects*. New York, Oxford University Press. 42

Schmitt, R. W. (1990). Engineering, Engineers, and Engineering Education in the Twenty-First Century. Belmont, Maryland, National Science Foundation and National Academy of Engineering. 33

Schneider, J., J. C. Lucena, and J. A. Leydens (2009). "Engineering to Help: The Value of Critique in Engineering Service." *IEEE Technology and Society Magazine* Vol. 28 No. 4. 40

Schumacher, E. F. (1973). *Small is beautiful: economics as if people mattered*. New York, Harper Torchbooks. 29

Seely, B. (1999). "The Other Re-engineering of Engineering Education, 1900-1965." *Journal of Engineering Education* **88**(33):285–294. 26

Smith, M. A. (2008). "Engineering Slavery: The U.S. Army Corps of Engineers and Slavery at Key West." *Florida Historical Quarterly* **86**(4):498–526. 16

Stokes, R. G. (2000). *Constructing Socialism: Technology and Change in East Germany, 1945-1990*. Baltimore, The Johns Hopkins University Press. 22

Strobel, J., I. Hua, F. Jun, and C. Harris (2009). *Students' Attitudes and Threshold Concepts Towards Engineering as an Environmental Career: Research by Participatory Design of an Educational Game*. "Sustaining Sustainable Design," Mudd Design Workshop VII, Claremont, CA. 38

United Nations Intellectual History Project (2005). Reflections on United Nations development ideas. Proceedings of the conference from development to international economic governance. Geneva, United Nations. 32

Velazquez, L. E., N. E. Munguia, and M. A. Romo (1999). "Education for sustainable development: the engineer of the 21st century." *European journal of engineering education* **24**(4):359–370. 38

Walker, P. K. (1981). *Engineers of independence: A documentary history of the army engineers in the American revolution, 1775-1783*. 16, 17

Williams, D. (2001). *Culture Wars in Brazil: The First Vargas Regime, 1930-1945*. Durham and London, Duke University Press. 18

Williamson, B. (2007). *Small scale technologies for the developing world: volunteers for international technical assistance, 1959-1971*. Society for the History of Technology, Washington, D.C., SHOT. 24

Winner, L. (1986). *The Whale and the Reactor*. Chicago, University of Chicago Press. 31

World Federation of Engineering Organisations (1997). "Commitment to Sustainable Development. Resolution adopted by the WFEO General Assembly." http://www.wfeo-comtech.org/ 36

Zwally, K. D. (1996). *Highlights of the recommendations of the President's Council on Sustainable Development*. Energy Conversion Engineering Conference (IECEC), Washington DC, IECEC. 35

CHAPTER 3

Why Design for Industry Will Not Work as Design for Community

"With over four decades of experience with appropriate technology in the South, why do so many engineering-for-development initiatives still struggle to produce successful, sustained outcomes? One compelling answer to this question is, simply, that 'Development is difficult.' This claim is a truism for anyone experienced in development work, surely. But if this truth were widely known, we might expect to see fewer projects initiated with more investment dedicated to each. Instead, the past decade has seen a proliferation of engineering-for development projects, and this fact provokes a different answer to the question of why engineering-for-development projects struggle with success: 'Our models for development are wrong.'"

–Nieusma and Riley, "Designs on Development: Engineering, Globalization, and Social Justice."

3.1   INTRODUCTION

We care about design. Of all engineering activities, it is perhaps the one creative process where science, math, art, economics, function, form (and more) can come together in the conception, development, and implementation of a system, or artifact for a specific purpose. Design is at the heart of what engineers do. We agree with Bill Wulf, former president of the National Academy of Engineering, who describes engineering as "design under constraints." After participating in design workshops, teaching an engineering design course, and conducting ethnographic work on engineering design activities at large high-tech companies like Airbus, Boeing, and Honeywell, we have come to appreciate the challenges that engineers face when teaching, learning, and doing design. In short, we celebrate design!

Yet after conducting numerous interviews with students and faculty involved in design for community development or humanitarian engineering—which we will call "design for community" for the remainder of this book—we became concerned about how the assumptions, methods, concepts, and practices underlying many of their design projects come from practices born in industrial and corporate settings. As we continued to listen to students and faculty, read their design reports and syllabi, review engineering education journals, and participate in design conferences, we further confirmed that most in engineering design education still follow assumptions, methods, concepts, and practices that come from industry and corporate settings.

While teaching courses on Engineering for Sustainable Community Development (see Chapter 8) and Humanitarian Engineering Ethics, many of our students involved in design for community projects found a place, perhaps for the first time, to critically reflect and write about the problems that emerged when they brought industry-based assumptions and practices to design for community projects. Students' realizations, and more importantly their capacity to grow intellectually as they made these realizations, were both revealing and inspirational to us (see Chapter 8 to see how students changed by questioning design projects and practices). As we did for our own students, we want to help engineering students involved in design for community make these realizations and grow intellectually in their own terms. Hence, we invite you to read this chapter and patiently engage the issues and questions raised here. In short, we have sketched here an anatomy of senior engineering design with the following goal: **To help students identify and question the underlying assumptions, concepts, methods, and practices in their engineering design courses and projects so they can assess the appropriateness of these for design for community.**

We begin our anatomy of design by dissecting a design project that won an award for "Exceptional Student Humanitarian Prize." Winners for this competition were selected "based on the results achieved and their impact on humanity, or, on a community." Although this is only one project of design for community, it represents an exemplar within the engineering profession, as it was sponsored by the president of one of the largest and most influential engineering societies in the world. We understand that every design project is different, so the project presented here might not be similar to yours. The features, assumptions, methods, and procedures in this exemplar might not be representative of all design for community projects out there. In other words, our intent here is not to generalize from a sample of one. Yet this exemplar is significant and relevant because of the recognition that it received from an engineering society, how it reminded us of the dozens of design for community projects that we have come across over the years, and how it can help you learn to raise critical questions on a completed project before you begin yours.

Our intention here is not to blame or embarrass a particular design team or course, but to begin revealing the hidden assumptions that are often made in design for community projects. Hence, the quotations that follow are taken verbatim from the source material, but actual references are omitted, so that the focus can be on highlighting oversights, inconsistencies, and challenges that such groups face when practicing design for community. Although lengthy, we decided to include here the majority of the project description, broken in segments, to allow you to slowly read, pause, and reflect through the all the steps taken by a design team, asking with us difficult questions that might reveal problematic assumptions, concepts, methods, and processes often made or used in design for community. Here we go!

3.2   ONE DESIGN PROJECT: DESCRIPTION AND REFLECTION

The description of this particular group's design project begins this way:

Project Description: In the developing world, there is a need for technologies that make their lives easier. These technologies need to be inexpensive and the materials must be locally available. One of the needs is an efficient and inexpensive way to crush and dehusk grain. The grain crusher project arose from an Engineers Without Borders trip to Senegal. They said that the people there do have a grain crusher, but some cannot afford it due to the cost of diesel, which is the fuel used to power the crusher. Also, the hand method of mortar and pestal is very time consuming and hard on the body....

This description clearly and neatly lays out a defined problem, one that perhaps you agree needs to be addressed. At first glance, it seems this design team is working toward something worthwhile and necessary. Yet, this description takes many things for granted:

- What assumptions exist when students call a country "developing" or "Third World," while calling others "developed" or "First World"?

- Why do engineers in these types of projects focus on "needs?" How do they find out what a community might "need"? How can these students come to assess and understand a community's needs and define a problem after just one trip to Senegal, despite language and cultural barriers between the students and the locals?

- How and why did local villagers express their desire to and interest in working with the design team? Is the students' assumption, that hand techniques are "time consuming" and "hard on the body," fully accurate? Is it a perspective shared by the local community?

In fact, the project description provides a narrative describing, in some detail, how the project members came to a solution:

Solution: The human powered grain crusher project began in the Fall of 2006. The original design concept was for a rotary stone grinder called a quern. *It was decided* that the most reasonable device to fulfill the purposes of this project would be the quern which is essentially two circular stones, one on top of the other, with an axle in the center and a handle attached to the top stone. The grain to be ground is placed between the two stones and the top stone is rotated about the axle. On more advanced designs, a small hole in the top stone allows for the continual introduction of fresh grain into the space between the stones. It is very efficient, effective, and constructed of all natural materials, which require very little machining. The quern and its direct descendent, the millstone, were so effective that they were the primary means for all grain production until the late 1800's. Some small scale modern mills, in fact, still operate using high quality, electrically-turned millstones that produce flour which is said to have better baking qualities than commercially available flour which is produced using metal grinding devices. *Having chosen a design,* using circular cement "paver" stones and a steel rod as an axle, a working quern was assembled and tested. *It was tested* by grinding various types of grains and proved to be adequate as a working model. Then the group started to contemplate what improvements could be made to a device that had existed in various forms for thousands of years. It was at this time that *the realization was made* that the quern had very little development potential, while keeping the cost at a reasonable level (italics added).

You probably detected that all of the italicized words in the description above are in the passive voice, which is a typical communication strategy in technical writing. Yet, the use of the passive voice here—in a design for community project wherein communication and collaboration between design team and community members is important—leads us to ask several questions:

- What are the implications of the passive voice ("it was decided", "having chosen a design", "it was tested") in a description of a design project?

- What does the passive voice hide? What does it tell us about how decisions in a design project are made?

- Might the passive voice reveal how the perspectives of the community that the design is supposed to serve remain untapped?

The project solution description continues:

Having realized that making a quern would not allow any improvements beyond what already existed, the focus of the project shifted to a grinder that is produced in Uganda. The Ewing III grinder is produced in a manufacturing plant in Uganda that was set up by Compatible Technology International (CTI), an organization that helps to improve food processing operations throughout Africa. Discovering the Ewing III grinder allowed us to shift our focus from designing a complete grinder to developing improved methods to power an existing grinder. We attempted to contact CTI in an attempt to acquire a Ewing III but never received a response. As an alternative, we selected the Country Living Grain Mill as a comparable substitute to the Ewing III grinder. For powering a grain crusher, a device is needed to convert human power to mechanical power for the grinder. Designs brainstormed and researched included bicycles or stationary bicycles modified with a chain or drive belt used to turn a crank on a personal, kitchen type grinder. **The group decided that a bicycle stand for an existing bicycle would be the best idea for the scope of the project.** A bicycle stand *was constructed* with intentions to be attached to a pre existing grinder. Of critical importance to the design was a wide range of adjustability so that the final product could fit a variety of bicycles. The stand would need to fit bikes with tire diameters ranging from 20 inches to 26 inches, and also with varying rear axle widths. The design also had to allow for adjustment to the tension in the drive belt, so *it was decided* that the grain crusher's location would be adjustable to provide such tension. The only fixed components would be the center drive axle and its supports. Everything but the bolts and bearings is made of 6061 aluminum, because this is just a prototype. The rear bike wheel is held in place by two "pucks" with holes lathed into them so that it fits over the nut on the back axle. The support shafts that hold that puck are adjustable in height. The wheel rests on a roller once it is properly secured. The roller has 80 grit grip tape on it to ensure more friction. When the bike is pedaled, the roller turns a 3" v-belt pulley which is belted to a 12" pulley on the grain crusher itself. Slots are milled into the base so that the belt can be tensioned or replaced. *An effort was also made* to use as many off-the-shelf pieces as possible. This would limit machining time and product variability for the end-user. *It was also proposed* to include an electric motor that could be powered by solar energy. This would give users the option of human or electric power, so if they do not have electricity, they are still able to use the device (italics and bold added).

Although the active voice (e.g., "we selected," "The group decided") is encouraging, active decisions seem to be made only by the design team and not by local community members. Again, questions arise:

- Could it be problematic to assume that a grinder produced in Uganda, or one sold by Country Living Grain Mill through a US catalog company, would be appropriate in Senegal? After all, these three countries have vast differences in people, colonial past, geography, economy, and potentially beliefs about, and ways of using technology.

- What issues might emerge when assuming that "off-the-shelf" parts found in the US would be found on shelves in a community in Senegal?

- How did the engineers know that the potential users would want to pedal a bike as a source of energy for the grinder? Who did they have in mind when selecting a bike? Women? Children? The elderly?

The project solution description continues:

During the Fall 2007 semester, the objective of this project returned to the original objective with a focus on reducing cost and weight of the product. In order to accomplish this, the Spring 07 design was turned into a working model and analyzed to determine how best to reduce the cost. It was found that if the team could create its own grain crusher instead of ordering the Country Living Mill, the cost of the overall product would be greatly reduced by about $150. This change also reduced the weight of the product because the crusher has been made from aluminum instead of much heavier cast iron. The new crusher is designed to retro fit onto the previous base. It is the same width and roughly the same height. The method of grinding remains the same. The grain is fed in through the top of the grinder and is augured through a hole. That hole allows the grain to fall between two grinding plates. One plate is bolted down, and the other rotates and the shear force is what grinds the grain. Keeping the same idea of the prototype in mind, the total price of materials was a large difficulty. The pricing goes hand in hand with the conservation of materials, so when the excess material is trimmed down, the total cost of the product will have the same effect. The Spring '07 design had a Baldor motor ($225), Grain Crusher ($375), and construction materials ($200), so the cost of this design was approximately $800 to build. For the Fall '07 design, the cost of the materials to make the crusher itself is approximately $224, which shows that the objective of this project was reached. The new design costs $151 less than the old one, excluding the motor. The motor was also eliminated from the design, saving more on the cost of the project. This would be one of the designs taken to India for the pilot study.

By now, you might be asking with us:

- Why are cost and weight reduction the guiding design constraints? How would the design be different if the main constraints were community empowerment and respect for traditional grinding practices? How was the design tested and by whom?

- What does it mean when the design cost of a grinder is $800 for a country with an annual GDP per capita of $1600?

- Why would this design, intended for a Senegalese community, be appropriate for testing in communities in India? Is this testing in India motivated by a belief in the *universality of technological applications* highlighted in the Introduction?

The design group goes on to describe the testing for the project, and its potential implementation:

This design was tested for how long it would take to grind different types of grain and roughly how much energy was required to grind each type of grain. A simple test was done by tying weights to a string and then tying the string to the shaft of the crusher. The maximum weight that we could lift at a replicable rotational speed of about 1888 RPM was 44.85 N. Since a replicable design was available, [our] University Business Department showed interested in working with the project. After meeting with the group of business students to show them our drawings and explaining the concept of the project they decided to work on the marketing side of the project. At present, the business team is doing research on areas the finished project can be marketed in. They are also presenting to a group of possible entrepreneurs that might be interested in becoming involved in the project. In addition, the design team also investigated other market sources for the product. During spring break, the Engineers Without Borders [EWB] clinic team traveled to a community in Senegal. Presented with this opportunity, a list of questions was compiled for the team to ask the community that they stayed with. A great deal of information was brought back from the group.

From this description, important questions emerge:

• Where were potential users from the community during this testing stage?

• How extensive, reliable, and legitimate could the data be if it was gathered during a one-week visit? Was this the first time the team engaged the community's perspective?

• Who defined the questions to be asked of the community? What did the team learn from the community?

The design team then describes some of the EWB traveling team's findings:

They found that there is already an existing grain crusher that can be used, but the cost of the diesel to run the crusher prohibits some of the communities from using the grain crusher. Also, the women of the community are the ones who crush the grain everyday, and to do so, they wake up around 4:30 am to produce enough grain for the day. The EWB team found that if an easier and less expensive way was available for crushing grain, the members of the community would be very interested. To provide visibility for the product being designed, a website was assembled to showcase the Grain Crusher project …It also gives links to the types of products used to create the grain crusher device. The site is currently up and running and has been updated recently with a photo gallery and video clip of the working model. In the beginning of the Fall 2008 semester, it was decided that the grain crusher assembly should be made to be more reliable and user-friendly. The idea behind redesigning is that the current design may not be acceptable to all bicycles, such as mountain bikes or bikes with pegs. Also, it was decided that the grain crusher could be made cheaper and lighter, but still be very sturdy by using steel. The new design still uses pedal power to turn the crusher, but it is self contained so that you will not need to attach a bicycle. To save money, the new design is made primarily out of steel. Steel is less expensive than aluminum, but it is stronger so fewer support components will be needed…It is also chain driven instead of belt driven. Another great benefit of this design is it not only reduced the cost of the crusher to $250 all together, the weight also dropped from about 40 lbs to 32 lbs. The design has the same grinder plates and loading system as the Fall 2007 design, except an auger was added to increase the feed rate of the grain that is being ground up. The seat has 6 adjustable height levels to make useable to a wide variety of people. The gear ratio is slightly greater than 1:1, but is still relatively easy to pedal. Energy calculations still have to be done on this design, but from observation the speed is slower, but it is easier to pedal than the Fall 2007 design. This design does have a slight learning curve when it comes to finding the correct height and getting used to pedaling while sitting directly above the axis of pedal rotation. The cross member that runs perpendicular to the chain direction is bent 5 degrees on both ends to reduce the "rock" effect and provide greater stability. With two designs completed, preparation began for the pilot study. One slight modification to newer design was made. In order to save money, each design was brought as luggage on the flight. The one end of the base had to be trimmed an inch and a half in order to fit the airline's luggage size regulation. This did not affect the balance or the performance of the crusher.

Salient questions emerge:

- What does it suggest about current design for community projects when students, *after* defining the problem and brainstorming solutions, find that the community already had a working grinder?

- If a key problem with the existing grinder is the cost of diesel, could engineers envision simpler *economic* solutions (e.g., subsidies or increased efficiency in diesel transportation, etc.)?

- What if most or all bicycles used in this community are needed for essential transportation between villages and are unlikely to be given up for grinding (see Figure 3.1)?

The description goes on to provide information about the village and the process of dehusking. Oddly, this information appears late in the description, long after the problem and solution have seemingly been decided upon, by the design team itself:

Sengalpaddai was the village in which the pilot study was conducted. The first part of the pilot study involved investigating, first hand, the current methods used to grind grain. One aspect that was not known was that they use a quern to "dehusk" lentils. Dehusking is the process in which the outer skin is removed and the lentil itself is broken in half. Once they separate the skin from the grain, they grind it up using a long stick and a bowl, a dry mortar, and pestle, or a wet mortar and pestle (for flour). The next step of the pilot study was to demonstrate how to use the Fall 2008 design. Once it was assembled, and a short demonstration was given, they took turns using the crusher. Once they had done that, we repeated the process for the Fall 2007 design. There were several major aspects that were observed. The first is that when adjusted properly, the crusher also removed the skin of the lentils very well. Also, not many of them wanted to sit on the seat and pedal, especially the women, because they did not want their clothes being caught up in the chain. Instead, they sat down behind

Figure 3.1: Health care workers in Senegal, Namibia, and other African nations use bikes to deliver food, medicine, and companionship to people with HIV/AIDS.
(Source: http://www.rd.com/content/printContent.do?contentId=58758 Author unknown.)

the crusher and pedaled by hand. Another noteworthy observation is that the Fall 2007 may be a little faster, but it is difficult to collect the crushed grain. Overall, the newer design was much better received than the older design. Once the trial was done, a roundtable discussion was had in order to receive feedback and suggestions to improve upon the design. They said that when they wanted the grain fine, it took too long so they wanted larger grinding plates to improve capacity. They also said that the grain tended to get caught up in the plates and just repeatedly cycling around, so they suggested making the plates horizontally oriented, in hopes of expelling the crushed grain faster. Another design change is to make it spin faster to increase output, but also keep in mind it needs to be easy to use. Another suggestion was to make one hand powered handle, instead of the foot pedals. They also wanted a better collection system.

Key questions continue to emerge:

- What does it mean for a community when a pilot design is demonstrated to them after their voices seem to have been excluded from all prior steps in the design process?

- What can we tell about the way engineering students are being trained in design when they find out at this stage of the design process that "not many of them [villagers] wanted to sit on the seat and pedal, especially the women, because they did not want their clothes being caught up in the chain"?

- How could a "roundtable discussion" yield trustworthy information when there has not been enough time or opportunities for the community to develop trust with the design team? How well-equipped are the engineers to discern the community's diversity of voices, hidden tensions, and conflict from a "roundtable discussion"?

The design team goes on to describe the current status of the project, and how team members envision future work:

> Current and Future Work: Currently, the newest design is being contemplated. The objective is to incorporate as many of the suggested design changes as possible. Price is still a parameter, but since only the design drawings will be sent to India, size and weight is not as large of a concern. One change that will be implemented is larger grinding plates. Madurai Mill Stores (www.maduraimillstores.com) sells much larger grinding plates. Currently, they have yet to send us technical information about the plates. Once the new design is completed, and a prototype built, performance calculations will be done. The energy calculations for the new design will make sure it still is relatively easy to grind the grain. Now that it is known what the villagers want in the design, the newest design should be the most ideal. If the prototype works well, the drawings will be sent the DHAN foundation where they will distribute the device to villages that need it in the effort to make them self sustaining.

Such a description leads to several questions:

- What does it say about our current practices in design for community when it took this long for students to readily write "now that it is known what the villagers want in the design"?

- Could the reception of the new prototype by the villagers in Sengalpaddai be affected by the quantity and quality of time the design team dedicated to establishing rapport and trust with them?

- What could be problematic about assuming that a prototype will be distributed to villages across cultures, adopted, and become sustainable in the long term?

The team briefly describes the impact of the project:

> Impact: By providing a better device for crushing grain, they can produce more of it for the community. Not only that, but the devices, once distributed, should provide entrepreneurial opportunities to the villages and people running them. This should alleviate poverty to an extent in the developing world and improve the quality of life. Once the grain crusher is finished, other prototypes will be developed that address other needs in the developing world. These prototypes should have the same result.

Summary Questions:

- What could possibly be wrong with a design meant to "alleviate poverty" in the "developing" world?

- What if good intentions are not enough to produce effective technologies?

- Instead of "helping" people, could technology end up reproducing inequalities such as those of development projects described elsewhere in this book?

3.3   DESIGN COURSES AND DESIGN INSTRUCTION

Perhaps, after wrestling with the questions above, you have become somewhat skeptical about design for community projects like this. Now we want to turn your attention to a design course, the site where projects like these are conceptualized, planned, developed, tested and written up, all activities for which students receive a grade. Furthermore, the design course is the place where most students come to learn and do design for the first time. By dissecting the constitutive elements of a design course and asking you to conduct some exercises along the way, we hope to facilitate critical reflection about the potential origins of the assumptions, methods, processes and concepts in many engineering design for community projects. After all, the problematic assumptions made by the student team above had to come from somewhere.

3.3.1   SYLLABUS

Syllabi are social contracts between professors and students, for they spell out what a course is about, what students are expected to learn and do, how they are going to be graded, what will be covered, and what is accepted behavior, writing guidelines, etc. Hence, syllabi constitute important evidence of how faculty understand a topic, how they are planning to teach it, and how they expect the students to learn it.

All engineering courses and their syllabi have at one point or another been justified in relationship to ABET accreditation criteria. Design courses are no exception. We found the following definition of engineering design from ABET in the course syllabus that we analyzed for this chapter:

> "…the process of devising a system, component, or process to meet desired *needs*. It is a decision-making process (often iterative), in which the *basic sciences, mathematics, and engineering sciences* are applied to convert resources optimally to meet a stated objective. Among *fundamental elements* of the design process are the establishment of objectives and criteria, synthesis, analysis, construction, testing, and evaluation." (italics added)

This definition reaffirms "needs," but does not explicitly mention the concerns or aspirations of those to be served by the technology. This view of need may implicitly reaffirm a problematic assumption that communities are deficient in something and that engineers are endowed with special knowledge and skills to come in and fulfill those needs (see Chapter 4). We invite you to question this assumption throughout the book.

This description also places higher value to knowledge coming from scientific analysis, testing, and evaluation. But what if you want to begin a design by seeking input from community members who have knowledge about their own locale and circumstances that might be considered non-scientific or too subjective? Is it possible to take them seriously and still be true to the definition? Also, how could this definition be appropriate for design for community given that, for example, it leaves out contributions that social sciences (such as anthropology), participatory techniques, or even aesthetics can make to designs responsive to community's concerns? This can help explain why the design project analyzed above did not include any of these perspectives throughout the research and development of the grinder. The fundamental elements of the design process in this definition do not include finding out community concerns and aspirations, defining problems with communities, or making iterative exchanges with community throughout the design process. Might this omission help explain why none of these important considerations were included in the design of the grinder?

3.3.2   OBJECTIVES

The syllabus that we analyzed for this chapter continues: "This course has been designed to comply with the ABET guidelines that require the engineering design component of a curriculum to include at least some of the following features…". The left column Table 3.1 lists these features, while the right column includes features not listed in the syllabus but that could be appropriate for design-for-community projects.

3.3.3   CONSTRAINTS

Likely your design faculty view engineering as "design under constraints." Hence, they might include a list of constraints that students must consider during their projects. The main constraints often included are the following: economic factors (e.g., time, cost), weight, safety, reliability, aesthetics, ethics, and social impact.

These constraints come to engineering design from a number of sources such as economic considerations, codes of engineering ethics, industrial practices, corporate values, and fear of liability. While these factors may be suited to industrial design projects, how appropriate is it to assume that these constraints easily apply or are equally relevant to design for community? We pose a number of questions here:

Table 3.1: Design objectives for industry and for community.	
Features actually listed in the syllabus	**Additional proposed features that might make the course appropriate for design for community (not listed in syllabus)**
"development of student creativity"	development of students' *empathy* for and understanding of a community's capacities
"use of open-ended problems"	defining problems with communities by *listening* to multiple community perspectives
"development and use of design methodology"	development and use of participatory practices and methods
"formulation of design problem statements and specifications"	formulation of circumstances that allow communities to articulate their own problems (concerns), specifications, and desires
"consideration of alternative solutions"	ensure that potential solutions include those generated by the community or by both community and team working collaboratively
"feasibility considerations" (often defined in terms of time, cost, weight, ease of manufacturing, legal and safety requirements)	allowing input from the community to weigh its own considerations and decide which are most important
"detailed system descriptions"	include a detailed socio-cultural description of the community to be served

- **First**, whose economy? What kind of economy? Budgets for design projects in the for-profit sector assume that everything can be bought and sold at a price (labor, tools, materials, land, permits, consulting, etc.) and that at the end there should be a profit. But what if a project is carried out in a different kind of economy where the most important assets defy simple quantification (e.g., local knowledges), others defy simple valuation (e.g., land), while others can be obtained for free (e.g., volunteer labor)? As one engineering design professor confessed to us,

  > [u]nfortunately, by the time engineering students are seniors–it seems as though they are totally convinced that what matters is cost, cost, and cost. This may be because cost is easily quantifiable—and therefore appears in many engineering problems given throughout their education…. Design is always about finding a solution that simultaneously satisfies (imperfectly) multiple goals at once.

  Design for community challenges engineers to consider multiple goals, many of which reside in the community and defy quantification. We understand that projects ultimately have to be economically feasible, but design-for-community projects require engineers to place community goals first. These goals may very well include making a profit, but would likely also include issues of justice and participation. At the end, the community should be the one establishing priorities.

- **Second,** whose safety? Safety depends heavily on cultural notions of "risk." What is an acceptable risk in one community might not be acceptable in others (Douglas and Wildavsky, 1983). Many water-related projects begin with the assumption that it is "risky," hence unsafe, for a community to drink water with certain levels of contaminants or bacteria. Quickly engineers assume that there is a "need" for cleaner water in that community, perhaps overlooking more serious community concerns. (See the case study in Chapter 6 to see what happened to a pre-defined water sanitation project when engineers began to listen to concerns instead of assuming needs). At the same time, different understandings of risk across cultural boundaries raise important, yet unexplored, ethical questions for engineers. Should an engineer accept a lower threshold of protection for a community willing to accept the additional risk?

- **Third,** reliability in technical systems is usually desirable. After all, it is great when things do not break often. But reliability comes at a cost. To ensure reliability, engineers might overspec a design or chose materials or parts with longer working lives but unavailable or too costly for a specific locality. A community might be willing to accept a significant degree of unreliability in exchange for local control over parts and maintenance. But those in charge of development projects seem to miss this lesson. After $53 billion of US taxpayer dollars spent in development reconstruction projects in Iraq between 2002-9, most of them failures, the US government is beginning to realize that high-tech or overspec projects may be less reliable and desirable than low-tech solutions (Adnan, D., 2009).

- **Fourth**, aesthetics present additional challenges to engineers working in "design-for-community"-projects. For US engineers, form and function are usually separate characteristics in the design of artifacts and systems. As engineering professor Louis Bucciarelli found out in his extensive ethnography titled *Designing Engineers*, "Though appearances can be central to the people specifically responsible for a product's looks, most participants in the design process worry primarily about how things 'perform' or 'behave.' Looks are secondary; function is primary. Designing in this sense is about how things work" (Bucciarelli, L., 1994, p. 1). Yet in many places outside the US or in fields such as fashion design, form and function are inseparable characteristics of an artifact, as lines, symbols, colors, textures, etc., have deep meanings and perform important cultural functions.

- **Fifth**, engineering ethics, usually contained in codes of ethics published by engineering societies, prescribe certain practices (what should be done) while proscribing others (what should not be done). For example, US engineers know too well that receiving gifts from a client could be easily interpreted as bribery. Most codes are very clear about this. But what if gift-giving is an essential act in building trust between a community and its outsiders who are coming to perform a service? In the book *Three Cups of Tea*, widely popular in the community development and humanitarian worlds, Greg Mortenson describes how important it was for him to receive gifts from a village before they will let him begin construction of a school. No gifts, no trust (Mortenson and Relin, 2006).

- **Sixth**, social impact, often treated in design courses as the "anything goes in here" constraint, is equally problematic, especially because often neither students nor faculty have the knowledge to understand and assess the social impacts of a technology. This kind of assessment requires education in areas like science and technology studies (STS), technology policy, or technology assessment, rarely found in engineering design courses (with few notable exceptions such as the Design, Innovation and Society Program at Rensselaer Polytechnic Institute). The hundreds of design projects that we have witnessed have never included long-term assessment of how the artifact or system in question impacts (for better or for worse) the socio-cultural environment in which it will live. Note that the project described at the beginning of this chapter did not include any kind of assessment of social impacts. In most written reports or final presentations that we have witnessed, students often make up something about the performance of the technology in the future, always portraying their design in the best possible light.

Exercise 8 *Can you develop a new list of design constraints, or modify the ones above, that might be more appropriate for design-for-community projects?*

3.3.4   EXPECTATIONS FROM STUDENTS

Like most syllabi, design syllabi include a set of expectations that faculty demand from students in order to give them a passing grade. These expectations are important because they socialize students to behaviors proper of most US engineering professional settings (e.g., corporations, government agencies). For example, the syllabi that we have studied for this chapter include *individual* expectations such as the following:

- "be on time and attend all class meetings, lectures, and team meetings;"

- "be respectful at all times of your teammates and faculty;"

- "maintain a well documented project notebook;"

- "submit professionally written progress reports;"

- "make professional oral presentations (each team member must present at least once during each semester);"

- "participate on all team assignments and functions (peer evaluations will be incorporated into the individual grade at the end of each semester);"

- "complete a one-on-one personal interview with the Team Project Manager;"

and *team* expectations such as:

- "formally document all client interactions (meetings, letters, memos, calls, emails, etc.);"

- "maintain a well documented team project notebook;"

- "demonstrate team effort on all team assignments and develop professionally organized oral presentations;"

- "work as a team toward a final product which is an effectively communicated team project proposal."

These expectations reflect professional norms and behaviors in corporate engineering settings. However, how do they relate to work with communities? We do not want to imply that communities deserve less respect from you, but we should remember that respect comes in different forms. For example, note the emphasis on "being on time," highly valued in US engineering settings. After all, in US settings, time is money. But how about in community development settings where "spending time" with and "being with" community members might be more important than "being on time" to a presentation? Learning to listen and building trust take time, an investment that defies quantification in terms of hours and dollars. Or how about communities where the pre-eminence of oral traditions precludes most attempts at documentation? Or where instead of presentations *directed to* a group (which might reaffirm assumptions about an expert talking to non-experts) what is valued is *conversations with* a group? In such instances, the above expectations seem to be misguided.

Exercise 9 *Develop a set of individual and team expectations that might be more appropriate for design projects in sustainable community development.*

3.4 COURSE CONTENT

3.4.1 DESIGN PROCESS

Engineering design courses introduce students to some form of design process. In most courses that we have looked at, the process is presented as a sequential model and visually represented as a step-by-step diagram that looks something like Figure 3.2.

Clearly, this is an elegant and clear depiction of design. It is hard to imagine, at first glance, what could be problematic with this model. However, engineering professor Bucciarelli describes the shortcomings of this *ideal* representation of design, and the possible motivation that faculty might have in continuing to reproduce this view of design in front of students, as follows:

> Such abstract figures express an ideal—an object-world creation of engineering faculty. Their intent is to *establish control* over the design process by breaking it down into discrete elements or subtasks, sharply bounding these subtasks by enclosing them in boxes or circles and then connecting them sequentially with straight lines. But while [this] figure and its kin may be useful pedagogically, in keeping with the reductionist tenor of such tools, as models of practical design activity they are deficient. If we allow the figure to direct our thinking about the people engaged in all the tasks contained in the boxes, we might conclude that design practice is an *extremely orderly, rational process* in which creative thought can be contained in a single box that yields a conceptual design or designs, which after detailed evaluation and analysis within some more boxes can be given real substance, tested, put into production, and then marketed for profit and the benefit of all humankind (italics added) (Bucciarelli, L., 1994, p. 111).

In his in-depth ethnography of engineering design, Bucciarelli actually found out that the design process is quite messy, far from orderly, and sometimes even irrational. After all, design is a human and social process of *negotiation and exchange*.

> To students, these diagrams shed very little light on how design acts are actually carried out or on who is responsible for each of the tasks within the various boxes. Nor is it apparent what these participants need know, what resources they must bring to their task, and, *most important, how they must work with others.* The lines with arrows hardly represent the *negotiation and exchange* that go on within designing…. As a reductionist, mythical, object-world representation, [this kind of] figure might be useful in the *indoctrination of students into the ways of thinking of the world of the firm*, but it misses the uncertainty and ambiguity of what really goes on in designing. Unlike the kinematics of particles, designing is not lawlike or deterministic. It is not a process of nature, nor can it be made to mimic nature (italics added) (Bucciarelli, L., 1994, p. 113).

As we will see in the chapters that follow, the processes of negotiation and exchange (of knowledge, skills, resources, etc.) when working with communities are perhaps more complex than when working with likeminded engineers within the same firm or from the same school. This may

Figure 1
Steps of the Engineering Design Process

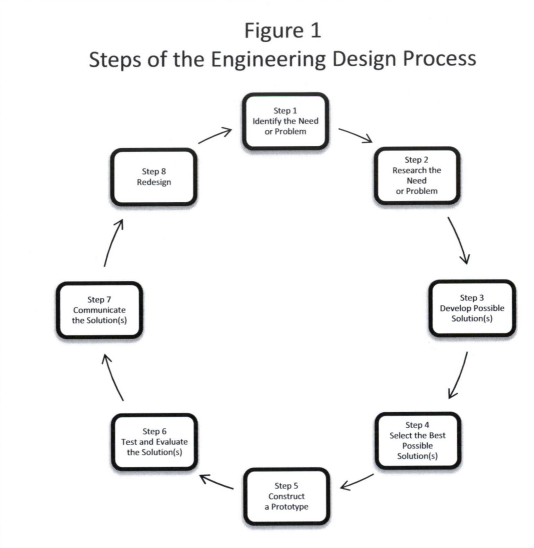

Figure 3.2: Sequential and step-by-step model of design found in many engineering design courses. (Source: `http://www.doe.mass.edu/frameworks/scitech/2001/standards/strand42.gif` Credit: Massachusetts Department of Education.)

explain in part why the description of the humanitarian design project above used the passive voice, hiding human agency, and social negotiation, and followed a step-by-step format that resembled the ideal model often presented to students. It seems that the students in that project were appropriately responding to a sequential model that they learned in class.

Exercise 10 *Share the sequential model and Bucciarelli's quotes above with your design faculty and discuss with them how these compare with what is taught in engineering design class. If this proves to be difficult, because you do not know your faculty well, share it with friends who have taken engineering design already. Invite them to think hard through their design experiences. How do they compare with the steps in the sequential model? What different groups of people or individuals emerged in each step? What kind of agreements and disagreements did they have? How did these human interactions and negotiations alter the course of the design? Try not to include "working with communities" just yet. Save that discussion for after you read Chapters 4 and 5.*

3.4.2   LEADERSHIP AND TEAMWORK DYNAMICS

Design courses often include presentations and readings on leadership and team work. Often engineering faculty (or visiting engineers from industry) in charge of those presentations have had extensive experience in corporate and/or military organizations. Given engineering's long historical association with the corporate and military sectors, this should come as no surprise (see Chapter 2). Hence, portrayals of leadership often include references to leaders in management/business and military (see Riley, D., 2008, Ch. 2 for a comprehensive analysis of how this historical association shapes engineers' "mindsets"). In some places, quotes on leadership often come from the works of management gurus like Peter Drucker and historical invocations of military leaders like Patton, Eisenhower, and Powell. Often in these presentations, students receive a portrayal of leadership as a desired set of characteristics of *individuals* who

- know how to take charge,

- move quickly to seize opportunities,

- have a vision and strive to have others share it as well,

- understand and respect their and other's strengths and limitations,

- listen more and talk less,

- and have passion for what they do.

As the story goes, design teams (or armies or business groups) with good leaders often succeed while those with mediocre leaders falter. The dynamic established is that between an *individual* who leads and a group of *individuals* who follow. Although there are variations on the treatment of leadership and teamwork from course to course and from school to school, the message here is clear: the human dimensions of design are centered in a leader who knows how to lead and a group that knows how to follow.

So what should be the appropriate approach when working with communities? Given the powerful message on leadership and teamwork, students might be left to conclude that the right thing to do is to "march on," seize a perceived problem in a community (e.g., their "need" for clean water), work as a group under effective leadership to solve the problem, and deliver a solution to the community. So what could possibly be wrong with that? As we will see in Chapters 4-7, this kind of approach has a long history of failures, especially under the umbrella of international development, and will likely fail again in design for community projects. Design for community requires a new kind of leadership (see Chapters 4 and 5).

Exercise 11 *Begin envisioning a type of leadership and team work required for engineering work in community development. After reading Chapters 4 and 5, put your own list of qualifications required of a leader working with communities in their own development.*

3.4.3   DESIGN TOOLS AND APPROACHES

Engineering design courses also include arrays of project management tools such as Work Breakdown Structure, Project Evaluation and Review Technique (PERT), and Critical Path Methods, to name a few. Among approaches to design, Quality Function Deployment (QFD) has gained popularity since the "quality movement" scaled up its incursion into US engineering practices since the mid-1980s. (Quality methods were one among many responses that US engineers put in place to address the challenge of Japanese technology in the 1980s). First developed in Japan in 1966, QFD was first used at the Bridgestone Tire Co. and Mitsubishi Heavy Industries "to identify each customer requirement (effect) and to identify the design substitute quality characteristics and process factors (causes) needed to control and measure it." According to the QFD Institute, QFD "was developed to bring [a close] personal interface to modern manufacturing and business. In today's industrial society, where the growing distance between producers and users is a concern, QFD links the needs of the customer (end user) with design, development, engineering, manufacturing, and service functions." QFD is now endorsed by a number of think-tanks dedicated to improve the quality of businesses (Mazur, G., 2009).

While we do not doubt the contributions that these tools and approaches have made to business and government in delivering higher-quality goods and services to customers, we wonder how appropriate these might be for design for community projects. What could possibly be problematic for community development about linking "customer needs" with engineering metrics or exposing trade offs between conflicting goals as QFD does? Well, all tools or methods developed for particular purposes and under specific circumstances (historical, political, economic) bring with them assumptions about the world around them and the people using them. Hence, tools developed for business bring with them assumptions (*inscriptions*) about markets where "customers" demand better products from "producers." In this world, customers exercise the power of their wallets (in the for-profit sector) or the power of their taxes and/or votes (in the public sector) to which producers try to respond with better products or services. But how appropriate is it to superimpose these assumptions on the realm of community development? What if the communities in question do not have the purchasing power to vote with their wallets or the citizen rights to vote with their votes? Furthermore, what issues are associated with treating community members as "customers" and engineering students as "quality experts" in community development? As we will see in the next few chapters, transferring assumptions from the business worlds (e.g., community = customers) is perhaps one of the most problematic issues confronting engineers in design for community projects.

Exercise 12 *Although we cannot analyze here every single topic covered in a senior design course, we invite you to think through other topics (safety, proposal writing, budgeting, reliability, ethics, intellectual property, etc.) and begin to question, to what extent are these topics appropriate for design for community projects?*

3.5   THE ACTUAL COURSE

3.5.1   LOCATION

Since design courses are scheduled like other courses, most design courses take place on campus in lab or conference rooms dedicated to design activities. Lecture presentations take place in large rooms or auditoriums with theatre-like set up. As the semester progresses, lectures begin to subside and classroom meetings might no longer be required. Often students move their project meetings to the library, an empty conference or classroom, or their own dorms. In any case, these settings are far removed from the community that is supposed to be served by the project. All designs are shaped by the environment in which they are conceived, the materials available nearby, and the colors, shapes, and textures around the design activity. All these factors influence designers in their choices and the outcomes of design (Rothschild and Cheng, 1999).

During a visit to the engineering design center of the French car manufacturer Renault, one of us witnessed how the entire facility was adorned with themes, pictures, colors and patterns from Brazil. Even the food and drinks served in the cafeteria were Brazilian. Renault engineers were getting ready to design a new car for the Brazilian market. When asked about this transformation of the design facilities, the engineer in charge of the tour explained that engineers/designers need to internalize, as much as possible, elements of the Brazilian landscape to able to generate designs that would appeal to Brazilians. They knew that the Renault facility would never be like the real Brazil, but at least the engineers understood the connection between design and place. So how can design for community projects be designed in a classroom or lab distant from the communities they are supposed to serve? As shown in Figure 3.2, important lessons can be learned from graphic designers who often surround themselves with artifacts, patterns, colors from the communities that they are intended to serve.

Exercise 13 *Survey the physical spaces where student groups are learning about, conceiving, and developing their designs for community. What do these environments tell you about the communities that the designs are supposed to serve? If travel to the community is impossible for you during the early stages of design, what could you possibly do to begin internalizing elements of the community and landscape where the design will eventually live?*

Figure 3.3: Design students at Drexel University Antoinette Westphal College of Media Arts and Design are challenged to design marketing materials to support the success of the Sunflower Oil Cooperative in the Ruggerero Genocide Survivors Village in Western Rwanda. Although no substitute for being in the actual place, students brainstorm over patterns, colors and textures of objects belonging to the community from which students derived inspiration. (Source: `http://blog.xcd.aiga.org/?p=502` Credit: AIGA XCD CrossCultural Junction Blog and Alan Jacobson).

3.5.2   COURSE PRACTICES

In spite of the problems with the sequential model of design, students in design courses quickly learn to follow these steps. After a number of introductory classes or activities, students move quickly to **problem definition** (steps 1 and 2 in the linear model discussed above). Through our interactions with students and faculty involved in design projects, we found out that most problems are defined *a priori* by others outside the communities to be served by the designs. These others often include professors, a representative of an NGO working near the community, a religious missionary, or sometimes the students themselves. In some complex cases, individuals or leaders in a community may make a decision without consulting others in their group. Such cases can be difficult to navigate. Reporting on how a design problem was defined for a design-for-community project, one student who analyzed one such project in a South American country said,

> [the professor] met [the leader of the ecotourism organization] at a sustainable resource conference in Boulder, Colorado. The two clicked and began working together. These two individuals laid the groundwork for the entire project, including all project goals

and constraints. [The leader of the ecotourism organization] was the sole representative of the students from [South American country], as was the professor for the students in [the US school].... A senior design student [from this project] observed that [the leader of the ecotourism organization] may have only been after personal recognition and achievement as [his organization] has won numerous awards (Note: names of people and specific countries were removed for confidentiality).

After problem definition, students quickly move to **choose design alternatives** that might solve the problem (step 3 in the sequential model above). Students often do lots of brainstorming on paper and/or computer software and search for parts and specs in hardware catalogs and/or stores available to them but perhaps unavailable to most communities that they are hoping to serve. Recall the selection of circular cement pavers, bicycles, and the Country Living Mill for the grinder design at beginning of this chapter. Unfortunately, students made problematic assumptions about the transfer of technology here. As the proponents of appropriate technology found out more than three decades ago, materials and parts for technologies to be used in "developing" communities need to be available (and ideally made) in the communities where the technologies will be used (Mason, K., 2001). If new technologies, dependent on parts and know-how from donor countries, are introduced (e.g., a water pump), these would likely disrupt community social relations such as those created by the women who often have the responsibility for collecting water. If the technologies eventually fail and cannot be repaired, the community ends up with both broken technologies and social relations.

In **selecting the best possible solution**, students follow the grade. If their faculty, either explicitly through the grading criteria or implicitly through conversations with students, emphasize *cost, weight, and timeliness* while undervaluing, if not completely neglecting, community empowerment, students would logically choose a design that reduces cost and weight and would be completed on time. Remember how in the grinder design above, the reduction of cost and weight became a priority for students during Fall 2007 even before they have checked their design with the community. We have yet to find grading criteria in a senior engineering design course that gives significant points for community empowerment.

Constructing and testing a design prototype (steps 5 and 6 in the sequential model above) are usually done in a lab or workshop on campus. Rightly so, faculty and students want to test a prototype under controlled conditions. After all, if the prototype fails in the lab, it will likely fail in the field. As one engineering design professor told us

Actually, I would encourage students to build and test prototypes in a lab setting before testing on the ground. Frankly, if the prototype doesn't work in the lab, under controlled conditions, it is highly unlikely to work in the field. Since engineering design is about using science and math to predict the performance of the design as it is developed—experimental data from the lab also is often a necessary supplement to theory. It is not and should not be the end result—and a prototype that functions in the lab may yet fail in the field—but this step should not be dismissed readily. It is a valuable part of the messy design process.

Yet questions for design for communities still remain. What are the limits of lab testing? The history of technology is replete with examples of how lab testing differs significantly from testing on the ground in the reliability, performance, and usefulness of a prototype. For example, engineers who developed a copper-cooled engine for GM in the 1920s found that the prototype performed well in the dynamometer but failed dramatically when tested in actual chassis and roads (Leslie, S., 1979). How much are students likely to get fixated on a design after prototyping and before they seek input from the community? How does prototyping in a lab reaffirm for students the belief that data coming from instruments are more dependable and reliable than input from the community? In the humanitarian design case study above, how might emphasis on testing and prototyping have obscured the fact that the Senegalese community had acquired extensive user knowledge from years of mill grinding experience?

Exercise 14 *Reflect on your interactions and practices of your design project. When and how might community input be included in multiple stages of the process, rather than simply at the very beginning or end?*

3.5.3   TEAMWORK

Often the division of labor among students in design for community projects is done according to engineering disciplines. As professors (or someone else besides members of a community) define the problem a priori, those involved in the management of student groups envision solutions and select team members according to engineering disciplines. Even those teams labeled as "interdisciplinary" or "multidisciplinary" are frequently made up only of engineering disciplines (Pierrakos et al., 2007).

A final report requirement in a design course actually challenges students to

- "segment the design into technical subsystems and implementation phases…"

- "assign responsibilities to each of the design team members for appropriate technical development and implementation milestones…"

- "include one-page resumes that highlight the design team members' technical capabilities corresponding to the Division of Responsibility."

With the grade at stake, students have little choice but to compose a team in this way. Also, since engineering design courses are usually available only to engineering students (with some noteworthy exceptions such as EPICS at Purdue or Stanford's Design Program), there are few opportunities for the inclusion of non-engineering perspectives. Even when faculty and students, either from engineering or non-engineering departments, want to include non-engineering perspectives in design projects, strict pre-requisites for engineering design courses get in the way.

Once the groups are formed according to an ideal engineering disciplinary mix, students become very pragmatic about this division of labor and what needs to get done. With serious time constraints and a grade on the line, they seek to maximize points earned, giving priority to the preparation and presentation of a final report, not to the people of a community. As one graduate student reported in an in-depth analysis of a design for community project, "[during the entire project] students' interactions with the village tend to be very brief and hurried, while their design, although brief as well, still receives the majority of their attention throughout the school year. This suggests that students, due to the limitations of the problem-solving-based curriculum and the overall time spent on design sans in situ, are *more likely to be aligned with the project than with the people*" (italics added).

Exercise 15 *If the priority of a design team is empowering a community by facilitating the solution of problems that the community defines on its own, how would you go about configuring a design team? What kind of talents, skills, and expertise are needed? Make sure to repeat this exercise after reading Chapters 4 and 5 of this book.*

3.6   THE WRITTEN REPORT

Students in design projects spend a great deal of time preparing and writing a written report for which they receive a substantial part of the course's grade. Although formats vary from place to place, here is a list of the sections that our students are expected to include in their reports:

- Letter of Transmittal

- Title Page

- Executive Summary

- Table of Contents and Lists of Figures

- Introduction

- Design Objectives

- Requirements, Constraints, and Criteria

- QFD Explanation & Summary

- Product, System, or Process Definition

- Deliverables

- Design Specifications

- Safety Analysis & FMEA

- Testing and/or Modeling procedures

- Division of Responsibility

- Project Schedule and Budget

- Qualifications

- Bibliography, Appendices, and Glossary

Although we could make an extensive analysis of format, style, and content, we will limit our observations here to a few questions:

- **First**, the format clearly reflects the sequential conception of design described above and reinforces corporate practices of communication and documentation. But how appropriate is this format to encompass the complex dimensions of designing and working with communities?

- **Second**, with very few exceptions, all the reports that we have read, including the project summary at the beginning of this chapter, are written in passive voice. Passive voice hides agency, hence precluding engineers' from responsibility and further decreasing the possibility to make community perspectives visible.

- **Third,** by appearing to be scientific and objective, in format and style, these reports limit their audience to those who value scientific writing and content. But how about those who do not, yet whose livelihoods might be directly affected by the design in the report?

Exercise 16 *If the priority of design for community is sustainable community development, how would you go about organizing reports of design for community projects? What kind of language would you use? What kind of sections would you deem essential to be included in a report? Repeat this exercise after reading Chapters 4 and 5.*

3.7   THE FINAL PRESENTATION

This is usually the concluding event for a project team's participation in a design course. It is an important ritual where students dress up, often wearing suits and dresses, compose impressive Power Point presentations, and often display professional behavior appropriate to corporate environments. Largely for practical reasons, teams present to faculty, students, and industrial clients but almost never to the communities that they are supposed to serve with their designs; hence, the questions and concerns raised during the presentations usually come from those in the audience (engineers), not from the community-recipients.

During presentations of designs for community development, students quite often present pictures of their brief visits to the community that give some degree of legitimacy to the team in relationship to their connection to the community (after all, they want to show that they went there). These pictures often include children smiling to and playing with team members, community members looking at the design prototype (perhaps for the first time ever), and students working on the installation of the artifact or system in question. Although well intentioned, these pictures hide and/or trivialize complex dimensions of community life. Rarely do these pictures show disagreements about the design, existing local technologies, knowledges, and practices that could have been considered instead of the design, or the long-term impact of the design in a community. Understandably, students are going after a good grade and want to show their design in the best light possible. Unfortunately, in the grading structures that we have found, there are no incentives for failures (although failure might be a greater teacher of design, (Petroski, H., 1985), controversies, or for acknowledging that existing technologies in the community might be more appropriate and effective than the students'. What would have happened to the grade of the humanitarian student project at the beginning of the chapter if from the onset the team had discovered and reported that the people of Sengalpaddai already had a technique to dehusk lentils?

Q&A sessions following the presentations often focus on technical details. In many cases, we have witnessed faculty, who in an effort to show that their students' projects are not fluff, ask highly technical questions, focusing on calculations, data reliability, and procedures. Although this questioning is the faculty's prerogative and responsibility, this behavior sends the message that technical details are more important than community matters. Perhaps this interaction is to be expected since there is rarely anyone with community-development expertise in these presentations.

Exercise 17 *If the priority of design for community projects is empowering communities, how would you go about presenting reports of designs for community development? What kind of format, language, and visuals would you use? What kind of interactions and audiences would you deem essential to be included in this presentation? Make sure to repeat this exercise after reading Chapters 4 and 5 of this book.*

3.8   CONCLUSIONS: WHAT CAN YOU DO?

Hopefully, this chapter has helped you identify and question underlying assumptions, concepts, methods, and practices in your engineering design courses, and projects so you can assess their appropriateness for design for community. Perhaps, you might be wondering "what now?" We would like to conclude this chapter with a number of recommendations that perhaps you can consider in order to reform those design practices that you have found problematic in your own educational context:

- **First**, take the exercises in this chapter as a starting point to begin sharpening your critical thinking about design for community. Try to complete these with project teammates and peers in order to elicit fruitful discussion and critical thinking about your own design practices.

- **Second,** begin implementing the lessons that make sense to you and your project. Whether you are about to begin a design-for-community project, or you are in the middle or end, there are plenty of opportunities to incorporate design-for-community lessons throughout. For example, if you are prototyping your design, you can acknowledge and report that the community has not provided input yet and hence the prototype has significant limitations. This acknowledgement is a good start because it puts the design in its proper place and reveals that the community's perspective, perhaps the most important, is still missing from the design.

- **Third,** constructively invite your faculty and peers to consider these questions and issues seriously. You might find resistance and skepticism but also more welcoming attitudes than you expect. We have found many engineering students and faculty that are ready and eager to begin reforming engineering design to make it meaningfully relevant to communities. Work with them in constructing alternative syllabi and new formats for written reports and final presentations.

- **Fourth,** take the questions and lessons from this chapter and book to sites outside the curriculum where engineers try to do design for community. For example, it is very likely that your school has at least one student organization dedicated to community development or humanitarian work (e.g., Engineers Without Borders, Engineers for a Sustainable World, Engineering World Health, Youth With a Mission, etc.). Engage them with the questions and issues that we have raised in this chapter.

REFERENCES

Adnan, D. (2009). U.S. Fears Iraqis Will Not Keep Up Rebuilt Projects, *New York Times*. NY. 68

Bucciarelli, L. L. (1994). *Designing Engineers*. Cambridge and London, The MIT Press. 69, 71

Douglas, M. and A. Wildavsky (1983). *Risk and Culture: An Essay on the Selection of Technological and Environmental Dangers*, Berkeley, University of California Press. 68

Dym, C. L. (2003). "Special Issue: Social Dimensions of Engineering Design." *International Journal of Engineering Education* **19**(1).

Leslie, S. (1979). "Charles F. Kettering and the Copper-Cooled Engine." *Technology and Culture* **20**(4). 78

Mason, K. (2001). *Brick by brick: Participatory technology development in brickmaking*. London, ITDG Publishing. 77

Mazur, G. (2009). "History of QFD." http://www.qfdi.org/what_is_qfd/history_of_qfd.htm 74

Mortenson, G. and D. O. Relin (2006). *Three cups of tea: One man's mission to promote peace …one school at a time*. New York, Penguin Books. 69

Petroski, H. (1985). *To engineer is human: the role of failure in successful design*. New York, N.Y., St. Martin's Press. 81

Pierrakos, O., M. Borrego, et al. (2007). *Assessment of Student's Learning Outcomes During Design Experiences: Empirical Evidence to Support Interdisciplinary Teams*. 4th WSEAS/IASME International Conference on Engineering Education, Agios Nikolaos, Crete Island, Greece. 78

Riley, D. (2008). *Engineering and social justice*, Morgan & Claypool. 73

Rothschild, J. and A. Cheng (1999). *Design and feminism: re-visioning spaces, places, and everyday things.* Piscataway, NJ, Rutgers University Press. 75

CHAPTER 4

Engineering with Community

As we explained in the Introduction, we could have chosen a number of titles for this book: Humanitarian Engineering, Engineering for Development, Engineering and Volunteerism, etc. There are so many of these programs popping up across the United States and in Europe that it can sometimes be difficult to keep track. Regardless of the term used, however, these many development-oriented engineering programs and initiatives have as their primary goal that engineering should be used to help those who are disadvantaged or in need. We settled on the title *Engineering and Sustainable Community Development* for this book because we thought it captured most closely what we value when it comes to these programs: sustaining the livelihoods and cultural capacities of communities involved in development projects. This focus echoes the definition of Sustainable Community Development (SCD) proposed by Bridger and Luloff, who argue that SCD must focus on "the importance of striking a balance between environmental concerns and development objectives while simultaneously enhancing local social relationships" (Bridger and Luloff, 1999, p. 381). This chapter is organized in a way that demonstrates how we arrived at that belief. The topics we consider here include:

- A definition of "community";

- **Challenge #1:** Engineering Problem Solving (EPS), and how it makes it difficult to put community at center;

- **Challenge #2:** Engineering mindsets, and how they can make it difficult to effectively consider community;

- **Challenge #3:** Curricular design (why most engineering for development is about *you* and not about communities);

- **Challenge #4:** Engineers' commitment to development;

- How to do things differently: A brief introduction to preparing for engineering *and* sustainable community development.

This chapter may challenge some core beliefs or motivations you have about engineering and its relationship to development, but our aim is not to be provocative without also being constructive. When we offer critiques of historical forms of development or specific engineering projects, we try to couple those critiques with suggestions for how to engage in this work more effectively, humanely, and sustainably. The truth is, although development projects as we know them have been underway for decades, the kind of work that we are engaging in now, as small groups of engineers, faculty, and engineering students, is new. We are learning together. Because of this, what we propose in this chapter is based on our best understanding of how and why particular projects have failed and how engineering can best move forward in working with communities.

Exercise 18 *Before reading the next section, write a one-sentence definition of community. In your sentence, address what community means in community development contexts.*

4.1   WHAT WE MEAN BY "COMMUNITY"

In Chapter 2, we presented evidence for the idea that "development" has emerged from specific historic and political contexts and as a result is a contested term: people disagree on what it means, whether it is beneficial, and about who benefits and who loses as a result of development policies (Vandersteen et al., 2009). The same could be said about the idea of "sustainability," also a very fluid and historically-situated concept, whose meaning changes depending on who you ask (e.g., McKenzie, S., 2004). By comparison, the term "community" may seem relatively straightforward: a *community* can be easily defined as a group of people bound together by geography, some common interest, history, characteristic or, in the case of many engineering projects, "need" or desire.

However, a closer look suggests that even for "community" there are multiple, competing definitions and meanings, depending on the context. And the definition you subscribe to probably reveals something about *your* values, worldview, and approaches, particularly when it comes to development work. Scholars who have been working in the field of community engagement for decades are still learning about and experimenting with definitions and methods for working with and understanding communities (e.g., see Burkey, S., 1993; Salmen, L., 1987; Salmen and Kane, 2006). As authors of this book we value definitions of community that are flexible enough to account for variable contexts, but that can also provide us with some guiding principles for engagement and reflection. As a result, we like the way that community-development practitioners Alison Mathie and Gordon Cunningham define community below, and propose it as a model engineers might consider when thinking about how to interact with communities they are working in or with. According to them, community is determined by the following:

1. **Relationships among its members**. Belonging to a community means being involved with the other members of that group in some way (Mathie and Cunningham, 2008, p. 7). This may seem obvious, but it's important to realize that the *nature* of these relationships can be highly variable. Relationships might be new and weak, as in the case of a group of people of different backgrounds coming together for the first time after a disaster (e.g., a tent city created after a hurricane) or old and strong, as in the case of a people from a village with ancestral attachments to each other. In either case, development projects should aim for respecting and strengthening these relationships.

2. **A relationship with place.** "Place" is loosely defined. Frequently, members of a community identify with a particular geographical place (like a village or city) where they are from or where

they live. But the place can also be virtual (like an online space, or a women's organization; Mathie and Cunningham, 2008, pp. 6–7). We argue that development projects should aim for respecting and strengthening this relationship to place.

3. **Differences in power and privilege.** These differences could vary in degree, from small—as when dictated by slight status difference—to very significant, as when shaped by a combination of socio-economic status, gender, race, and caste. In any event, development projects should aim for respecting these differences even when they might seem to go against Western ideals of equality. When a particular subgroup of the community appears to be oppressed, it is not the role of the Western "expert" to relieve them of this oppression but rather to facilitate their seeking alternatives if the members of the subgroups desire to do so (see also Guijt and Shah, 1998/2001, p. 8; Chambers, R., 1997, pp. 162–187).

4. **Alliances with a common purpose or purposes**. Communities may come together for a variety of reasons, whether for commerce, kinship, entertainment, or political cause. The rate of participation in these purposes may vary, depending on the needs and desires of individuals involved (Mathie and Cunningham, 2008, p. 7). Development projects should aim for awareness and understanding of these purposes.

The four characteristics listed above may seem aggravatingly abstract, obvious, or broad. Or, at first glance, they may seem to have nothing to do with engineering. And even if they do, how does one go about designing a project that respects variable concepts of interpersonal relationships, places, privileges, and purposes? In the abstract, such a definition can seem daunting. We provide a brief case study to illustrate, however, that it is often the engineer's ability to understand and work with the malleable, fluid nature of community that can make or break community development projects.

4.1.1   HOW ONE ENGINEER VIEWS COMMUNITY

In "The Stranger's Eyes," anthropologist Joyce Carlson tells of an enthusiastic young North American, Pierre, who had designed a new kind of mill for grinding grain. He believed his new mill design would make life easier for poor African villagers, especially for the women in those villages, who spent much of their days in difficult, back-breaking labor, such as collecting water. His project was funded by a church group, and he proceeded to Mali, Africa, to install his mills; how his project turned out is described below. Excerpts adapted from the original case as written by Carlson are presented here.

Exercise 19 *As you read "The Stranger's Eyes," note in the margins where Pierre, the young engineer, encounters problems with understanding 1) relationships among the community members; 2) relationships the community members have to place; 3) differences in power and privilege; and 4) community members' alliances with particular common purpose(s). When you've finished reading, try to deduce what definition or characteristics of community Pierre held before embarking on his engineering development project.*

The Stranger's Eyes

By Joyce Carlson

Supyire proverb: "The stranger's eyes are wide open, but he does not see anything."

In May of 1987, an aid organization with religious backing (ORB) sent a young North American named Pierre to Mali to install mills for grinding grain in various villages. Pierre presented his project first to the Association des Eglises Evangeliques in Bamako, which channels money into Christian aid projects throughout Mali. They suggested that he try to place his mills in the Sikasso region, traditionally the territory of the Christian Missionary Alliance. The CMA already had its own Aid Projects Coordinator. However, ORB did not intend to work through existing missions, so Pierre made no attempt to contact the CMA or to research the needs of the Sikasso region.

It was, however, his intention to work through local churches and through the women for whom the mills were primarily designed as a way to relieve their truly crushing workload, and to give them a sense of value, purpose, hope, and so forth. The women in this society are decidedly overworked. Before daybreak and long after sundown they are gathering firewood, fetching water, farming in the wet season, spinning cotton in the dry season, pounding grain for *toh* (millet paste), or making *beurre de karite* (cocoa butter).

From the time of marriage or before, the ones who are not sterile are either pregnant or nursing all of the time. A woman not carrying a baby in one way or another is considered unfortunate. All of them are exhausted, and many are anemic.

Kafinare already has a mill, right beside the paved road next to the market. It has been in operation since 1985, through the private enterprise of Ali Sanogo and his brothers, who are all residents of the village. How is the existing mill doing? Ali says that just to stay in business, he is having to charge 25 CFA less per customer than the nearest mill in another market town (15 kilometers away). During the dry season, he says, women like to save their money for more interesting purchases than getting their grain ground. During the rainy season, business improves because women are working all day in the fields and welcome a chance to skip the daily pounding. But the hard fact is, the existing mill is barely breaking even.

Pierre came as a stranger to Kafinare, asking no questions. It was nearly a month before he realized that he was putting a mill directly across the road from the existing one. In true Kafinarian fashion, no one told him that we already had one because he had not asked. When the truth eventually dawned, he protested in some shock that he would never have dreamed of running the enterprising villagers out of business, but then he plunged ahead with the plans on ORB's drawing board.

ORB sent Pierre to Mali with certain conditions attached to their gifts—conditions that mystified the villagers and made ordinary life a great deal more complex. The first was that no men should be involved in the ownership of these mills. This was a project for women, born out of a Western, feminist agenda. Secondly, Christian and non-Christian women were to collaborate in the administration of the mill and share in the profits. Thirdly, 10 percent of all profits from the mill would be given to the local church as a tithe.

To fulfill Condition 1, Pierre came to Kafinare, as he went to all the other villages that had been chosen to receive a mill, and called a general meeting of the Association Des Femmes (Women's Association). Problem: there was no Association Des Femmes. But, according to the agenda espoused by ORB, there must be, so an association needed to be created. The women of the three quarters of Kafinare were invited to meet at the church, and Pierre, who had been in Mali all of two weeks and fluently spoke his own French language, communicated with a Bambara translator. So during the first meeting, the women were brought to realize that they must elect officers so that their association would have an acceptable structure and would represent the wishes of the majority, and so forth.

From the women's point of view, there were problems with this first condition. They did not like it that the men were not to be involved. There are, of course, a variety of situations in this village where men are most unwelcome and definitely not invited. This situation apparently, as the women saw it, was not one of them. Immediately, the women were asking themselves, if the mill breaks down, who will fix it? Who will make sure that it has diesel to run? Who will really take care of the money? (Nobody in this village wants to be treasurer of anything because they will become a target for thieves and automatically come under suspicion if anything goes wrong).

So the women were already unhappy and anxious before the idea of the mill project was a day old. The men were insulted. It is traditionally their place to fiddle about with machines, such as plows and looms, so this prohibition gave the whole project an air of frivolity.

The second condition, that Christians and non-Christians were to collaborate, specified more precisely that Christian women were to be elected to the positions of president and treasurer in the newly formed association, and non-Christian women were to be elected as vice-president and secretary. Aside from the fact that surely this should not have been the business of ORB, far removed as they were, this condition took no account whatever of village norms.

When the elections were held, Pierre was thrilled to discover that the women were unanimous in all of their choices. They elected the oldest (and heaviest) Christian woman as president. Actually, the very oldest woman in the village should have been president, but, unfortunately, she was not a Christian and, consequently, did not meet the criteria for president. But she was elected vice-president. She has never come to a single meeting. The position of secretary (also designated for a non-Christian) went to Pauline, a school teacher and the wife of the school director. She was a Catholic, which for the Protestant majority was as good as not being a Christian at all. Most importantly, she could write, which was an important consideration, so she swallowed her resentment at being classified as a non-Christian and agreed to take the post. The position of treasurer went to Nema, the wife of Dioume, who was the head nurse in Kafinare. Why? Not because she was so good at counting, but because Dioume was known to be incorruptible, so the money would be safest at his house.

[There is a] notion of consensus that tends to hold sway in African villages. In reaching a consensus, people will talk over all the angles of a problem, and, eventually, they will come to a broad agreement that, if not doing away with opposing views, at least sweeps the leftover disagreements under the rug of mutual peace and goodwill. This consensus, however, at least in the case of the Supyire, remains on the whole a surface phenomenon. Discord, jealousy, and hatred remain alive and well underneath the rug of peace and goodwill. In the case of the mill, the village was split along the lines of those who were related closely to the owners of the already existing mill and those who were not. The anger and jealousy became apparent to me in conversations with villagers the day after the vote. Pierre, however, had already left the village, and when I told him about it later, he refused to believe me.

At the first meeting, all the women present were asked to pay 500 CFA into a common fund. Oddly enough, this free mill was going to cost the local community (including the men) both time and money. (The men of course were going to donate their time by making bricks, carrying stone, and putting up the cement structure to house it).

The mills were intended to be in place within three months of Pierre's arrival. In fact, it took a full year for the mills to actually begin operation. Pierre bravely weathered disappointment, illness, storm, and heat. He called the women together time after time and watched them force responsibility on the now reluctant men. With dismay, he saw money disappear like water into dry ground.

The Kafinare mill began operation two weeks before he was to leave Mali for what he hoped was forever. It was a great day of celebration. Everyone who came to the Grand Opening of the Mill would get one grinding free. (Some women who had paid 500 CFA a year before, were annoyed to discover that the initial sum would not entitle them to free grinds for the rest of their lives.) In all the time that the mill ran, it was rare to see the operators at the mill before 9:00 A.M., by which time Ali and Company had taken care of most of the business.

I went to Bamako the week that Pierre was leaving. At that point, three out of the four mills he had installed in the Sikasso region had broken. The grinding stones were meant to grind coffee, not grain. One week after he left, the Kafinare millstone broke. What was everyone to do now, since, of course, Pierre had gone, and ORB was, as far as we know, not planning to send anyone else? Pierre had looked ahead to such a contingency, and he had gotten the Compagnie Malienne de Textiles (CMDT) to promise to fix anything that ever went wrong with the mill. Unfortunately, he never got this agreement in writing, and the CMDT has, in fact, never lifted a finger to help repair the mill.

Source: http://www.sil.org/anthro/articles/thestrangerseyes.htm.

There are a number of ways to view this case. From one limited perspective, the project can be seen as a success: Pierre had a clear plan in mind, and he worked diligently to see the project built. His intentions—to help alleviate some of the difficulties he felt women in the community endured—were good. He considered the needs of the villagers as *he* imagined them and planned for the future of the project by involving community members in its establishment: he even made arrangements for future repairs.

From another perspective, this project was very much a failure. The mill may not have been necessary at all. Pierre seemed to leave the project with all of his flawed assumptions about the community and culture intact, and he may go on to execute similar projects, wasting good will, time, and resources. Ominously, it could also be argued that the community was left not better but worse off after this "development" project.

Exercise 20 *Now that you have read "The Stranger's Eyes," outline what Pierre, the young engineer, could have done to respect and enhance the following: 1) relationships among the community members; 2) relationships the community members have to place; 3) differences in power and privilege among community members (and between the community members, Pierre, and the organization he worked for); and 4) community members' alliances with particular common purpose(s).*

Clearly, though based on a real experience, this is a teaching story. We might think about how things could have turned out differently had Pierre involved the Kafinare citizens in a discussion about their lives and how his engineering knowledge could be of use to them (if at all). Following this kind of discussion, it is possible that the mill would never have been built; it is possible that another "problem" altogether would have been identified; it is possible that the young man's expertise would not have been necessary or useful at all. Equally important, the engineer might have learned new techniques, approaches, or solutions to problems had he decided to engage the community and listen to how they dealt with the issue of grinding grain in the past. This alternative approach—rooted in participatory development and community engagement, rather than in engineering problem solving—might have proven frustrating for Pierre, who wanted to solve problems for "the poor."

But, despite these possible frustrations, the argument of this chapter is that *community engagement* is also a more ethical, sustainable and effective way of approaching engineering development projects. We outline some guidelines to help engineers prepare for engineering development projects, and in Chapter 5, we provide specific guidance for community engagement through listening. Before moving to these approaches, however, we aim to provide a clear-eyed analysis of the difficulties engineers face in making community a central part of engineering for development.

Exercise 21 *Outline the obstacles that Pierre encountered in his project. How many of these obstacles did he create? How many were outside his control?*

Before we go on to discuss some opportunities we see for making community a central part of engineering development projects, it makes sense to lay out some of the obstacles to effective community engagement, even when your intentions for community involvement are good. Awareness of these challenges is an important first step toward increased self-awareness and better project planning and outcomes. Here we describe four specific challenges: 1) engineering problem solving, 2) engineering mindsets, 3) curricular design, and 4) engineers' belief in development.

4.2   CHALLENGE #1: ENGINEERING PROBLEM-SOLVING (EPS)

Although we must avoid making blanket statements about all engineering development programs, our research suggests that these programs are not always engaging meaningfully with the notion of "community" or with approaches to community engagement. In other words, the case of Pierre and the Kafinare is not an isolated incident, but it is representative of how many engineering development projects turn out. As engineer Donna Riley notes, even for small-scale development projects, "Assessment throughout a project, and particularly years out, is often poorly executed or nonexistent. Critical thinking about adverse impacts or the social context and cultural impacts of projects is often lacking" (Riley, D., 2007, p. 4).

Our own research on engineering for development projects bears this out (Schneider et al., 2009). We think that one of the main reasons that engineers and engineering students struggle with incorporating community into their development projects is that they are deeply invested in an approach called Engineering Problem Solving (EPS). Many students or faculty may not realize that this is only *one approach among many* to identifying and solving problems because it is so deeply ingrained in how they have been educated. It may seem almost natural, as if no other or better alternatives exist. But we believe that EPS, as a strategy or method for approaching community development projects, has significant disadvantages.

US engineering students learn EPS throughout their basic science and engineering science courses. Presented in Figure 4.1, EPS includes six steps—Given, Find, Make Assumptions, Diagram, Equations, Solution—and is often presented to students as a methodical, rational way of identifying and solving problems. In some cases, particularly when students are working on close-ended textbook problems, EPS is an effective method. In courses ranging from Physics to Thermodynamics, students are required to use EPS thousands of times by the time they graduate (Downey and Lucena, 2006). Furthermore, engineers and engineering students may see it as a value-neutral way to tackle problems and as an approach that can be applied to nearly any situation. As we noted in Chapter 1, however, EPS is *not* neutral with respect to incorporating a diversity of perspectives.

You may already be able to guess that dividing the world into people who define problems in a "right" way and those who define them in a "wrong" way can have significant implications for engineers who want to work in development contexts, with diverse communities. Excellent engineering students may, according to Downey and Lucena, "emerge from engineering curricula knowing that engineering problems have either right or wrong answers […]. In the process, they have acquired solid grounds, seemingly mathematical, not to trust the perspectives of [those] who define problems differently (Downey and Lucena, 2006).

Exercise 22 *Think back to Pierre from the case study above. In what ways did he employ an extension of EPS? In which ways was he given a problem? Made his own assumptions without checking with the community? Made diagrams and worked through equations without any input from those he was supposed to serve? And came up with one solution that he thought was clearly right? What were the advantages of his using this approach? What disadvantages or blind spots did it create?*

We don't mean to suggest that EPS is unimportant or flawed for all contexts, nor are we implying it can't be used effectively in some development contexts. Rather, we suggest that it should be one approach among many that engineers undertake when working with communities on development projects—it should be *a* tool for problem-solving, rather than *the* tool. Furthermore, students need

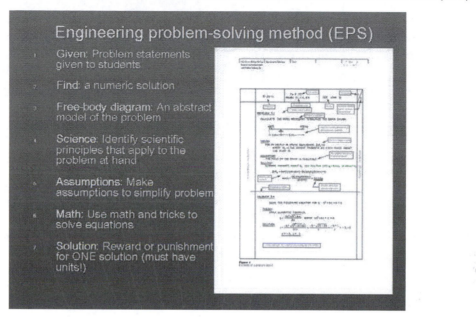

Figure 4.1: Engineering Problem Solving as presented in engineering science textbooks. The format clearly communicates to students that there are well-defined boundaries between the technical problem and the social world and an orderly sequence of steps involving math and science (and little of anything else) that must be followed in order to find the one true answer to the problem. (Source: Juan Lucena).

to be aware that their education in EPS may have pre-disposed them to identify non-engineering perspectives as "wrong" (or at least suspect), while valuing their own as "right."

Bernard Amadei, founder of EWB-USA, and William Wallace, author of *Becoming Part of the Solution: The Engineer's Guide to Sustainable Development*, note that

> The challenge is for the engineering profession to develop a new repertoire of appropriate technologies and associated best practices. We need best practices that can be used at different community scales ranging from large cities to slums and refugee camps. We also need best practices that can be used on different time scales such as those in the prevention, rapid response, recovery, and development phases of disasters associated with human and natural hazards. Such a range of best practices does not exist (Amadei and Wallace, 2009, p. 11).

You may be wondering where this leaves you as a student interested in engineering development work. If EPS is the dominant method in your engineering education but has significant limitations akin to those encountered by Pierre in the Kafinare story, and if "best practices" for development work haven't yet been developed and institutionalized in engineering education, how are you to proceed?

Our answer? With openness. With reflection. And by listening carefully to those you will work with, and who have done this kind of work before you. We think of this as "expansive" or "holistic" thinking, which encourages multiplication of perspectives and approaches to problem definition rather than reduction of them. We provide some useful practices in the next chapter for you to consider, but we would also invite you to ask yourself the following questions about your project, which are adapted from those posed by Caroline Baillie in her book *Engineers within a Local and Global Society*:

- Rather than using a cost-benefit approach, ask yourself *who* is bearing the costs of your project, and *who* benefits?

- Does the project promote justice? Does it distribute the costs, risks, nuances and benefits equally among all those involved in the project?

- Does it restore reciprocity (i.e., does it allow for an *exchange* of some sort between you and the community, or among community members, or with others)?

- Does it promote conservation over waste?

- Does it favor reversible over irreversible outcomes? (Baillie, C., 2006, referring to the work of Ursula Franklin).

Exercise 23 *Using the list of questions above, return to the Kafinare story "The Stranger's Eyes." Imagine that you are Pierre before the project has begun. Ask yourself the questions listed above about the project. Which questions do you have answers to? Which do you need to find out more about? How might thinking in EPS terms lead you to define terms like costs, benefits, and justice? Conversely, how would thinking holistically or expansively change your definitions of these things?*

4.3   CHALLENGE #2: ENGINEERING MINDSETS

It should be clearer now that communities often resist being thought of or treated as "problems" to be solved. They may resist boundaries such as those that engineers draw between the technical and social dimensions of a problem. They resist quantification, especially when their cultural frameworks are rooted in non-scientistic ideas. For US engineers, math is the language of nature. For a community, the language of nature might be math imbued in mythology, rituals, or the qualitative nature of the stories passed from one generation to the next. Or, they may have quantitative methods that differ from those used in EPS. It can be maddeningly confusing to the outsider (or, we hope, exciting and inspiring), and quite time-consuming to understand and make sense of nature, humans, and the relationship between them. These frustrations can be exacerbated if an engineer approaches problem solving with a "planner's" mentality, with efficiency as priority, and with an over-confidence that one's technologies and "solutions" for the community are best. Not everyone shares faith in the power of technology, especially when that technology has not been contextualized within a community's worldview, location, knowledge, and desires.

In her book *Engineering and Social Justice*, engineer Donna Riley writes that engineers frequently fall into particular mindsets as a result of their education and workplace environments. Although engineers themselves can be a diverse bunch, we can nonetheless identify some key mindsets that might define them as a group. Many of these are related, though not exclusively so, to the EPS approach we defined above. They include the following:

1. A strong commitment to problem-solving using an EPS approach, often ignoring context and values.

2. A reliance on the scientific method as the primary way of knowing the world, to the exclusion of other ways of knowing.

3. An intense focus on work and achievement, with a "narrow technical focus, perhaps including as well a denial or devaluing of relationships and enjoyment."

4. A commitment to militaristic or industrial work contexts (though sometimes paired with a critical awareness of this commitment).

5. A commitment to "uncompromising objectivity" (Riley, D., 2008, pp. 33–45).

These are mindsets that engineers are themselves frequently aware of, argues Riley, and that may lead to undesirable outcomes. Particularly in the case of work oriented toward social justice concerns—such as engineering development work—such mindsets can prove to be a real hindrance to engineers (Riley, D., 2008, pp. 43–44).

Exercise 24 *Think about your own way of living in the world in specific terms, from how you drive your car to how you make purchasing decisions to how you interact with your friends and professors. In what ways have you internalized EPS or the mindsets that Riley lists above? How have these mindsets served you in positive ways? How might they make things challenging for you or those around you?*

How might they shape interpersonal relationships with people, groups, or communities who hold different ways of knowing, seeing, and being? Can you imagine ways of doing things differently that might be equally fulfilling or effective? If you have resisted these mindsets in particular contexts, can you explain why?

Many of the engineering mindsets described above have, in fact, been historically problematic for community development projects. We identify two principal problems: 1) engineers' commitment to a "planning" mindset in development projects and 2) limitations created by engineers' commitments to "help."

Engineering as planning. As was explained in Chapter 2, engineers have been part of colonialist histories that have sought to control peoples through the use of "scientistic" approaches that have had lingering and sometimes devastating effects. Large-scale development projects have been particularly prone to failure, from a historical perspective (see Chapter 2). Economist William Easterly suggests that the reason for these failures is located in a "planning" approach to development. Planners, argues Easterly, use top-down methods and treat poverty as if it were "a technical engineering problem that [planners'] answers will solve" (Easterly, W., 2006, p. 6). Planners are interested in "grand utopian Plans that don't work" (Easterly, W., 2006, p. 369). Such plans usually attempt to address many problems at once, are devised by planners unfamiliar with realities on the ground, and cannot be evaluated in the long run.

Key Terms

Planners: For the purposes of this book, we think of Planners as those who design development plans based on boundaried, quantifiable information. Planners may have some knowledge of what is going on "on the ground," but are typically interested in applying plans in a top-down fashion, regardless of local circumstances.

Searchers: We think of Searchers as those who situate themselves within communities, and who endeavor to identify problems *with* community members and seek practical, local, and/or holistic solutions to those problems. They are adaptive and typically unwedded to grand plans.

Engineers have a long history as planners (see Chapter 2) and the "planner" model of development is, as Easterly notes, reminiscent of EPS, in which problems are carefully bounded, segmented, and quantified. One point we aim to develop in this chapter is that engineering education is structured to be particularly well-suited to the "planning" sort of development that Easterly critiques. It is a top-down, systematic, analytical way of thinking about solving problems in which community processes and communication are often made secondary or invisible.

"Searching," on the other hand, poses challenges to engineers who have been trained to approach problems in one particular way. Searchers, argues Easterly,

> can look for piecemeal, gradual improvements in the lives of the poor, in the working of foreign aid, in the working of private markets, and in the actions of Western governments that affect the Rest [those not in the 'North…']. Searchers can gradually figure out how the poor can give *more* feedback to *more* accountable agents on what *they* know and what *they* most want and need (Easterly, W., 2006, p. 30).

In other words, searchers by necessity undertake numerous strategies to understand the community, including actively seeking out feedback from multiple stakeholders and putting in place mechanisms to make diverse stakeholders accountable for their development results. As we described above, the "searching" approach may be exactly what engineering students are educated to turn *away* from.

Exercise 25   *Read the article below by Nalini Chhetri. As you read, note which portions of the story reflect a planning approach and which are more representative of searching. Where is EPS used? Where are more holistic or open ways of interacting and problem solving used? How might things have gone differently if, instead of planning, the engineers had searched?*

Excerpt

Excerpt from "Lessons from Ghana: Why Some Technological Fixes Work and Others Don't" By Nalini Chhetri

This summer, while interacting with villagers in the Western African nation of Ghana, along with a team of faculty and students from ASU, I was able to test [the idea] that a successful innovation (policy) rests upon the wisdom to know which problem will cede to technological solutions and which ones will not. ASU's engineering faculty and students, involved with [an NGO], had developed technological fixes to solve two perceived problems of smoke from firewood cooking, and unreliable electricity in Ghana – they were stoves that operated on smokeless gelled ethanol and "Twig Light" technology, respectively. We needed to find out which ones would be embraced by villagers and which ones would not be, and why? We also needed to know if problems of villagers as perceived by our engineers and students paralleled those of the villagers themselves. In short, were we developing technological fixes for real village problems or for problems that *we* perceived to exist?

Last year, as part of ASU's student capstone project, facilities to produce ethanol, a smokeless fuel derived from corn, had already been developed, shipped and installed in the village of Domeabra. Engineering students from Kwame Nkrumah University of Science and Technology (KNUST), a local university, had declared this facility to be operational. The next step was to come up with stove that would operate on ethanol that would be gelled.

This spring, a group of male and female engineering students designed stove prototypes that catered to a family of five. The final product was a round, insulated, approximately 20-inch-high single burner device with a fuel tray insert on the side and a flat top. It was smokeless, odorless, clean, and efficient. The stove itself was significantly more efficient than a similar prototype that had been developed in South Africa, and the gelled ethanol was about five times cheaper than the ones being marketed in Accra, the capital of Ghana. However, ASU students would find out that their stove had several design flaws, considering families used round bottomed cooking pots that required rigorous stirring while preparing their main meal and used intermittent high heat. The cost of the fuel, too, was high, because villagers used free firewood. Through a mapping exercise we conducted, we found out that a typical home had anywhere from 10 to 21 family members, a far cry from the small family that they had envisioned. Also smoke while cooking was not considered a problem, rather a part of daily life.

The Twig Light prototype was a different story. A simple lighting device built as a low cost alternate light source for poor villagers, the Twig Light was made from a low cost thermoelectric generator using twigs to light a bank of LED lights that was enough to light up a small room. When it was demonstrated for the first time, villagers suggested using some burning embers/charcoals from the fireplace instead of twigs and *voilá* – the LED lights lit up bright enough to read a book in a dark room. We told the villagers the Twig Light generated 5 volts of electricity, and an immediate comment was – could it also be modified to recharge cell phones?! It had become obvious to us that many villagers that were with us had cell phones.

A lot of the homes in the villages had electricity, but it was unreliable and expensive. In some cases, electricity had been cut off as villagers were unable to pay the steep monthly bills. So a notion of a lighting device that needed no batteries, was low cost, easily accessible, and could generate electricity fascinated villagers. When the demonstration was over, determined groups of women passionately requested to be allowed to have these devices immediately. When we reminded them that they were prototypes and needed to be refined, they thought we were playing hard to get, and they offered to buy them on the spot! That it took a lot of persuasion to make them back off is an understatement.

So what problems yielded to technological fixes and what did not?

The reasons could potentially be found in the fact that the Twig Light may work as this technology addresses what Sarewitz and Nelson refer to as the [embodiment] of "cause-effect relationship connecting problem to solution." In other words, the low cost and simple Twig Light solved a clear problem of high cost, unreliable, centrally controlled electric service. Infrastructure development that provides basic services in developing nations are notoriously dysfunctional, but electricity that brings light in homes is clearly a felt need that has met with success in improving people's quality of life, while allowing them to keep their values and interests, lifestyles and habits.

The gelled ethanol, on the other hand, marketed to address the problem of deforestation and health issues were indifferently dismissed by the villagers. For them, firewood was free and easily accessible from their farm plots. Smoke was part of daily cooking – they did not perceive it as a health problem. Adopting the improved stoves would mean changing their lifestyle, habits, values, and interests. So they were not ready to fix a problem they felt did not exist for them.

This article originally appeared as a Center for Science and Policy Outcomes (CSPO) "Soapbox" post. Nalini Chhetri is a postdoctoral research associate at CSPO, and a lecturer in ASU's School of Letters and Sciences (Chhetri, N., 2009).

See the original post at `http://www.cspo.org/soapbox/view/090827P6KF/lessons-from-ghana-why-some-technological-fixes-work-and-others-dont/`

Exercise 26 *In this account of "Lessons from Ghana," what engineering mindsets are at work? What actions tend to reveal a Planner mindset? A Searcher mindset? How are these at work at various project phases, such as problem conceptualization and definition, exploration of possible solutions, and project implementation?*

Engineering as helping. For engineering professionals and students who are drawn to engineering development work because of a desire to help or to serve, engineering for development programs can offer what seems like an entirely new framework for doing engineering work by helping the underserved. Engineering for development's core values—altruism, sustainability, justice—are noble, seemingly beyond reproach. Yet, there are bodies of scholarship that have for many years interrogated and critiqued the history of programs like these, and which have lessons to offer engineers. It is important to broadly outline these critiques.

First, engineering development practitioners must understand that their projects have emerged from or are merging with the history of development, as we described in Chapter 2. Although engineering educators seem to be aiming their efforts *away* from this history (of exploitation and inequality) when they develop programs devoted to engineering for community development, they may be in danger of entering into and repeating problematic development practices (Epprecht, M., 2004; Jackson, J., 2005; Rist, G., 2002; Sichel, B., 2006).

Such a possibility exists primarily because engineering development programs are fundamentally based on an inherently problematic model of "they" need/"we" help. Having emerged out of the history of development and conceptualized through the lens of engineering problem solving, engineering for community development is still committed to a paradigm that imagines the "developing world" as characterized primarily by *needs* (Riley, D., 2008; Selingo, J., 2006). This is problematic because *engineers often interpret needs as parameters or constraints in their problem solving or designs*.[1] The concern is that the more engineers conceptualize their relationship with communities or the "underserved" in terms of need/help, the more they see communities as defined by what they lack, while re-affirming themselves as "problem-solvers" or "planners" with solutions (Easterly, W., 2006; Schneider et al., 2008; Sichel, B., 2006).

4.4   CHALLENGE #3: COMMUNITY DEVELOPMENT PROJECTS

This challenge can be briefly summarized this way: educational engineering development projects are designed with you, the student, *not community*, in mind.

The fact is, most of the engineering for community development projects we are examining in this book are developed as a part of *your* education. Faculty believe that these projects will help you learn how to apply engineering knowledge in a real-world context, expose you to global issues, teach you about communication, and so on. In some cases, this is exactly what happens. One could look at any number of engineering development programs and see small groups of engineers and engineering students coming together, "searching" for small-scale problems that engineering know-how might solve. But we believe these projects raise some important questions that must be addressed: 1) Do engineering development projects teach students important lessons about key concepts such as community, sustainability, and communication? and 2) How do we know that students are not

---

[1] Maslow's hierarchy of needs provides engineers a convenient framework to reduce and homogenize human needs, regardless of cultural diversity, to five categories: physiological, safety, love/belonging, esteem, and self-actualization. Engineering "solutions" tend to focus on the physiological (air, food, water, shelter, excretion) and on safety, when the context allows, but rarely address the other three categories (Maslow, A., 1943).

Figure 4.2: An African woman receives food aid from the West. The temptation is to see her as emblematic of a third-world person "in need," as opposed to asking complex questions about her country's debt relationship to international organizations such as the International Monetary Fund; the social or environmental reasons she is receiving aid to begin with; and the desires or motivations that lead to her accepting the aid.
(Source: `http://trendsupdates.com/the-truth-about-third-world/`)

the only beneficiaries of these projects (or, to put it another way, how do we know that communities have been truly served)?

Within engineering education, engineering development projects typically involve teams of students from Canada, the United States, or Europe, occasionally in partnership with students from host countries. These teams try to design small-scale, "appropriate," ecologically sustainable, and inexpensive technologies for use in communities in developing countries, or in "underserved" communities closer to home. For example, Figure 4.3 shows students from Duke University in the US visiting a developing-world hospital because they want to send free spinal-injury devices to patients there as part of an innovation competition at the university. Frequently part of design classes or senior capstone projects, such efforts may involve long-distance communication between students and "clients"—a community spokesperson or NGO contact—in the host community. Sometimes they involve a brief one- to six-week trip to the community, for the purposes of fostering cross-cultural communication between students and community members, gathering information for designs, or installing technologies. Most often, these projects involve no travel at all due to cost, length, conflict with academic semester or safety concerns associated with international travel.

Figure 4.3: Duke University Engineering Students visit a developing-world hospital. (Source: http://prattpress.pratt.duke.edu/duke_spine Credit: William Richardson.)

Our own research shows that few engineering for development programs require any education in the relationship between technical and non-technical elements of such design projects (see also Mulder, K., 2006). Nonetheless, most of the professors working in these fields argue that such programs have positive outcomes for engineering students, though little long-range or in-depth studies have been done to examine what it is, exactly, that you as students are getting from these projects.

On the one hand, we believe design for community projects may help effectively prepare you to work in areas that are of increasing importance in today's world, such as environmentally sustainable engineering and design. And, like engineering professor George Catalano, we agree that engineers have an ethical obligation to redress poverty, inequality, and environmental damage: "While engineering is a profession with a strong ethical dimension... there has been until very recently no reference to addressing two of the most important issues of our times—poverty and underdevelopment and environmental degradation.... [I] believe that we, as engineers, need to change the way we envisage our profession" (Catalano, G., 2007, p. 2). Design for community projects have the potential to sensitize engineering students to these significant concerns and reflections.

On the other hand, design for community projects have the potential to affirm problematic student beliefs or perceptions. We describe four of these potential pitfalls here:

1. **Who benefits and who pays?** There may be an imbalance of benefits accrued in design for community projects. As was the case with colonialism and international development, our students and engineers from the developed world have much to gain from these projects (experience, travel, knowledge, and skills which can later translate into better job opportunities) and almost nothing to lose. The communities involved *may* gain useful technologies, or may not, depending on whether the project works as expected (the Kafinare story illustrates this). But they also risk disruption, project failure, and loss or degradation of resources and self-determination as a result of interaction with engineering development projects.

2. **Stereotypes:** Without adequate preparation and reflection, students from rich countries in the project may have a number of their pre- and misconceptions about development and the "developing" world confirmed. As noted in Chapter 2, development projects often involve engineers from the Northern hemisphere attempting to assist communities in the Southern hemisphere. Development studies scholar Mark Epprecht asks thought-provoking questions of all work-study programs abroad: "…how serious is the risk that unexamined good intentions and high ideals could backfire …by actually hardening Northern students' pre-existing negative or exotic stereotypes about the South, by fostering a missionary zeal that alienates the wider public audience in the North from a critical understanding of North-South relations, or by creating in the Southern hosts feelings of burden or exploitation by the North?" (Epprecht, M., 2004, p. 689). Might community development projects in your school be contributing to creating or reinforcing stereotypes of how you see people from poor countries?

3. **Resentment:** The community may also suffer from interactions with students from rich countries. Again, Epprecht: "[I]t is not uncommon to find communities in the South…that attach unrealistic hopes to the arrival of helpful strangers from abroad. Yet, the stranger moves on and the heartfelt promises…are forgotten. Where most in the community may simply slough off the disappointment, for others it could feed into the very cultures of cynicism or victimization that international exchanges are supposed to assuage" (Epprecht, M., 2004, p. 694). Is this a possibility in design for community projects at your school? How will you know?

4. **Lack of context:** Design for community projects may affirm students' mistaken belief that large social inequities can be solved by "band-aid" solutions: "Band-aid solutions and little make-work projects may thus obscure the brutal reality of the international system as it stands" (Epprecht, M., 2004, p. 699). Engineering students may be particularly disadvantaged when it comes to being able to understand the forces of neoliberalism and globalization that frequently create or exacerbate the inequities engineering for development tries to address. Their involvement in development projects—if not set within this larger context—may confirm that narrow view of the world. How can you educate yourself about things like globalization and neo-liberalism? How might this affect the way you approach problem-solving in a community?

Exercise 27 *Return to the Kafinare story at the beginning of this chapter one final time. In what way did Pierre suffer from each of these four problematic beliefs or perceptions? Provide an example from the story* *to illustrate each. Then, propose one possible remedy for each: what are some ways that Pierre might have avoided or learned from these negative outcomes?*

4.5   CHALLENGE #4: ENGINEERS' BELIEF IN DEVELOPMENT

We argue that the idea of "community" should be made central to engineering development projects for this reason: if community (the people who are supposed to be served by development projects) are not meaningfully and respectfully considered and included in engineering development project planning, implementation, and assessment, it seems quite likely that SCD projects could run into a number of the same problems large-scale development "tragedies" have run into. Although it is not possible to re-create the history of development in this chapter, we think it is important to emphasize some of the major points about development made in Chapter 2 and how problematic it is for community development that many engineers still hold dear to the premises of development and modernization.

We argued in the introduction to this book that development is a paradox. Often, those of us from rich countries rarely question the wisdom or correctness of development. We believe so strongly in the narrative of development—of continual, progressive, civilizing human improvement towards higher levels of material well-being—that it has come to seem natural, as if there could be no other way of being in the world. Gilbert Rist details the history of how this came to be. In the 19th century, colonialism was justified by the use of the biological metaphors of growth and progress. Empires forced colonies, through violence or economic dependency, to join the bandwagon of progress, for the former argued that this was natural, unavoidable, and right. The principles of colonization, and later development, were seen as having "unchallengeable legitimacy. By making colonization out to be 'natural', it was possible to disguise the political decisions and economic interests lying behind it" (Rist, G., 2002, p. 54).

Since the mid-1950s, people from rich countries have held to similar beliefs about development being "natural" and "unchallengeable." It is right, we believe, not only to want clean air and water, healthy food, safe dwellings, education, and fulfilling and consistent livelihoods, but also to want access to abundant consumer goods and services. And we believe that it is our duty, in the North's rich countries, to extend a helping hand to those in the South, to help them "develop" (for more on the "global south" and "global north" see Chapter 2). Engineers tend to hold dearly to these beliefs. After all, they play a central role in the design, construction and maintenance of the technological systems that deliver these goods and services.

On the other hand, those projects that have been completed in the name of development— often huge projects intended to provide poor countries with services like clean water, electricity or schools—have by many accounts failed (see Scott, J., 1998, for an exhaustive account of how certain schemes to improve the human condition have failed). In fact, some projects, such as those that we introduce in this chapter and the next, have left people worse off than they were before the project began. There are those who argue that development has never really been about human growth, progress, or improvement at all, but rather is about "the general transformation and destruction of the natural environment and of social relations. Its aim is to increase the production of commodities (goods and services) geared, by way of exchange, to effective demand" Rist, G., 2002, p. 13 (see also Sachs, W., 1992). In short, the generally unstated outcome of development has been to make the rich countries richer while keeping poor countries poor.

This portrayal of development may seem cynical and overly economic; it does not take into account the common beliefs that those in the North are doing the right thing by "developing" other nations, and that Northerners should share wealth and can help. Yet, we observe our own historical moment and see that the world's biggest problems—poverty, climate change, access to clean drinking water, disease—persist and, in some cases, seem to have worsened in the decades since development began in earnest. These conditions stand in stark contrast to the relative wealth of developed countries. So we must wonder if Rist is right about the primarily economic and unjust thrust of development. And perhaps more importantly, these conditions may call into question the common beliefs in the premises of development.

Development, as a historical enterprise, has been spectacularly unsuccessful, and even disastrous in some cases. Shiv Visvanathan, an anthropologist and human rights researcher from India, writes "that a wailing wall should be instituted in the World Bank office just so that we could mourn or grieve together in the aftermath of some projects" (Visvanathan, S., 2005, p. 84). This is the central paradox of development, then; in rich countries, we have a belief in development that seems as natural and essential to our identities as breathing air, but which when spread to other parts of the world can seem like poison. It is important to acknowledge and understand that others may see development as poison, and not as a mutually beneficial enterprise.

We want to reiterate, however, that even given the great challenges involved in design for community—the required awareness of political, historical, and cultural context—we believe there are still ways to interact with communities that can be beneficial to those involved. The potential pitfalls are great, but we retain hope that with continued self-reflection and awareness, design for community projects hold promise.

Exercise 28 *Make a list of your interests in or motivations for participating in SCD projects. Reflect on this list. How many of the items you've listed have to do with your belief in the inherent benefits of development? How many assume that you have skills or knowledge to bestow on others? What has led you to those beliefs, and how might they be partial or even incorrect?*

4.6   FROM ENGINEERING *FOR* DEVELOPMENT TO ESCD

What the preceding discussion about challenges tells us is that engineers need much preparation before they embark on development projects. Such preparation is particularly important to understanding how what is considered "technical" is shaped by engineers' engagement with community and how community could be impacted by technology. Furthermore, we agree with Epprecht, who argues that students returning from the field also need assistance processing and understanding what they learned abroad.

We realize that as students, you may have minimal say in how a particular project or assignment is designed (although in Chapter 3 we gave you some critical tools to begin challenging how design for industry is currently applied to design for community). But you very likely have some control over how you interact with a particular community, how you go about educating yourself about cultural, political, and social issues related to that community, and how you reflect on your own motivations, involvement, and project outcomes. In order to facilitate these reflections, we provide below a list of concepts and strategies you can try to integrate into your own engineering development experience. Not all will be applicable to every situation, but as a whole, they may open up a more expansive worldview or mindset than those you've developed elsewhere in your engineering education. Some are about the "doing" of engineering and sustainable community development ESCD; others have more to do with understanding and reflection.

A few words of caution before proceeding. We want to make clear that we are *not* offering a simple formula here: sustainability + development + a pinch of community = Successful Engineering Development Project!. Rather, our aim is for engineers to develop tools for self-reflection and accountability as they enter into ESCD work. As Epprecht puts it, "The task, therefore, is to find a balance between presumption (of intrinsic good) and pre-occupation (with risks)" (Epprecht, M., 2004, p. 704). We know that it would be counterproductive for you to be paralyzed by a fear of doing harm or misunderstanding; at some point, you have to develop an understanding by engaging in the project itself, learning from mistakes and successes.

We also acknowledge that adopting a "blueprint" for community engagement may lead to a "planner" type of development, which ignores local contexts and lacks flexibility. As Guijt and Shah note, the "standardization of approaches...contradicts one of the original aims [of participatory development], to move away from the limitations of blueprint planning and implementation towards more flexible and context-specific methodologies" Guijt and Shah, 1998/2001, p. 5 (see also Cooke et al., 2001 and Hickey and Mohan, 2004 for an extended debate on the "tyranny" of participation). Instead, we prefer open-ended approaches to community engagement, such as that of contextual listening (see Chapter 5).

Finally, we want to warn that it is possible that women or minority groups or voices will be excluded from community engagement activities (see Crawley, H., 1998/2001). Significant work—such as that pictured in Figure 4.4—has been done working specifically with women on development projects, and it is important to pay attention to the outcomes and lessons learned from those projects. Development workers may need extensive time in the "field," with the host community, before they can begin to understand or grasp the dynamics in place—think again about the case of Pierre in Kafinare (see also Epprecht, M., 2004; Heron, B., 2007). Community engagement exercises, such as group meetings, may be slow, deliberate, and inconclusive, such that projects do not get completed. Furthermore, there may be varying interests competing for funding or particular projects—NGOs or particular community leaders may have notions of what should be done, and these may diverge from what the majority of the community wants (e.g., Page, B., 2003). Again, "community" is never homogeneous or monolithic.

Figure 4.4: Women in Mali meet at an Entrepreneur's Meeting. The meeting was organized by Oxfam as part of a "Sisters on the Planet" program.
(Source: http://www.treehugger.com/files/2009/10/women-most-renewable-resource-oxfam.php Credit: Rebecca Blackwell for Oxfam America).

In other words, engaging community is not a fool-proof solution to the problems of development work or ESCD. We acknowledge again that this may prove frustrating to some engineers involved in ESCD projects. Some may think communities should be grateful for their hard work and expertise, and should trust them and their knowledge. Understandably, they may also feel pressures from funders, universities, course instructors, or parents who expect to see projects completed in timely, efficient, and cost-effective ways. Engaging "community" could challenge all of these pressures and expectations.

That said, here is a list (though not exhaustive) of concepts, guidelines, and/or practices you may want to consider as you prepare your ESCD work. The list is followed by more extensive discussion of each item.

1. Incorporate opportunities for *self-reflection* before and throughout the project.

2. Find meaningful ways to *learn about the community* you are working with—their history, their language, their values—and get help with community engagement processes.

3. Figure out ways the *time-scale* of your project can be expanded.

4. Make *plans for "failure."*

5. Design a *landing pad for yourself*, and develop meaningful *assessments* of yourself and your project.

4.6.1   THE IMPORTANCE OF SELF-REFLECTION

Students should be involved in discussions about their (and their peers') motives for getting involved in this work, and they should think about the many situations, motivations, desires, and wants that have converged to bring them to this moment in their educations or careers. They might read provocative essays on development Illich (Illich, I., 1968/1990) or Gustavo Esteva (Esteva, G., 1992), and ask themselves the following questions:

- How do engineers or students see themselves in relation to the community they are visiting?

- How might the community see them?

- Are they there to provide expertise? To build something? To "make life better for poor people"? To earn a grade? To learn from the community? Or for other reasons?

- What might it mean to shift from focusing on the project to focusing on the process of communication? From talking to listening? From donor to receiver?

- What might it mean if an intended project is never finished? Or never begun?

When we teach our course "Engineering and Sustainable Community Development" (see Chapter 8), we encourage students to think about a number of issues related to these questions. For example, one aim of the course is to help students see the limitations in the "universalistic" notion that technology can be transferred from one context to any other without regard for socio-cultural, political, economic, and other dimensions that inform and are informed by community identity, values, and aspirations. At the same time, we encourage students to critique appropriate technology, the notion that simply because a technology is "small" or environmentally sound, it will be uncritically adopted by a community (see Chapters 3 and 8).

As much as it is important to know that there are methods and approaches for community engagement, perhaps even more important is to be reflexive about the attitude one brings to development work. We warn of the perils of approaching development engineering motivated by a desire to "help." The notion of helping frequently implies that a community is deficient in some way. Envisioning and listing all the elements that a community does not have, and going no further, emanates from a deficiency mindset, one focused on what is lacking. Instead, a capacity mindset asks different questions:

- What is present in the community?

- What capacities exist already in the community?

- What can the community teach me?

- What work might they be doing on their own "development?" (see also Chambers, R., 1997; Mathie and Cunningham, 2008).

As the participatory guidelines above clearly suggest, there are many assets (knowledge, expertise, leadership, innovative organizational arrangements, etc.) within the community that should be understood, valued and, if the community desires, used towards the successful implementation of ESCD projects. Your role might become that of a facilitator instead of helper.

Similarly, we believe that ESCD engineers will need to remain flexible and humble. They will need to understand that they may not be in control of situations (solving problems in their own terms as they learn in school), and at the same time, they have enormous power to do good or wreak havoc. Again, by observing the participatory guidelines above, engineers relinquish control of a project and foster community self-determination, enabling a community to deploy their own assets and be deeply involved in every decision, including those that might terminate a project.

Easterly argues that development workers would do well to "discard your patronizing confidence that you know how to solve other people's problems better than they do" (Easterly, W., 2006, p. 368). And grassroots journalist Ben Sichel puts it this way:

> … if I'm going somewhere to "help people," what am I assuming about those people? How effective a worker can I be in a place where I hardly speak the local language, and where my specific skill set might bear only a passing resemblance to what is actually needed? And—one of the most difficult questions of all—do the people I'm working with really want me there? (Sichel, B., 2006).

The idea that humility may be taught perhaps sounds strange. But we think that the self-reflection exercises recommended by Epprecht (see above), careful reflection on the history of engineers in development (see Chapter 2 of this book), and the listening approaches we outline in Chapter 5 are good starting points.

Although we would advocate that any student involved in ESCD work in the field also be required to take courses in development and on the area where the work is located, at the very least, we would like to see ESCD students guided to certain forms of reflection and questioning. Following Epprecht, we believe students need tools for self-reflection (whether in the form of journaling, group discussions, or other activities) to critically situate and reflect on their own role as development officials. As students, you can make these reflective moments part of your design processes, and incorporate them into your final reports. Invite your faculty to consider these reflections as part of the grade so even in a small way you can begin changing what is valued in your courses. These *reflexive interventions* should not be busy work: you should read and reflect on development case studies, chapters from this book or others like it, and/or texts on the country or community you will be visiting.

4.6.2   LEARN ABOUT THE COMMUNITY

Ideally, ESCD students and practitioners would take a range of courses to prepare them for their work in the field. For example, students from the US who partner with an NGO in Thailand to purify drinking water would have access to courses on Southeast Asian history, culture, and/or politics. Or they might have taken a course like one we offer at the Colorado School of Mines, that includes the history of engineers and development (see Chapter 8 for more). In the best of all possible worlds, they would have taken some Thai language classes.

We realize that this scenario is probably unlikely, however. Much more likely is that as engineering students, you *may* get some sort of general course in dealing with other cultures or a brief seminar on "culture shock." Perhaps you will only have a few preparatory meetings to discuss these topics before you leave. In your design courses, most likely one of you will draft a paragraph on "cultural impacts" for a final report.

But we want to challenge you to go beyond this. We challenge you to make learning about and from the community an integral part of your entire experience, whether it is volunteering or designing.

Exercise 29 *Think about the grain crusher project examined in Chapter 3, and about the Kafinare case above. How can you avoid falling into the same pitfall these engineers did? How can you find a way to communicate with the community members, to learn about them, to begin to see and understand their lives and motivations, way before your volunteer assignment or design project begin? How can you involve community members in your design from beginning to end? How will you assess your work? Or better yet, how can community members assess your work and find the opportunity to tell you when you are no longer needed?*

These questions require that you do far more than perform a cursory internet search on a particular part of the world (though that is a good starting point). But the benefits of exploring such questions to your overall project could be substantial.

For those involved in development work over the last decade, community participation and engagement have emerged as key approaches in small-scale development work. Land and water engineer Irene Guijt and rural development economist Meera Kaul Shah write,

> The broad aim of participatory development is to increase the involvement of socially and economically marginalized people in decision-making over their own lives. The assumption is that participatory approaches empower local people with the skills and confidence to analyse their situation, reach consensus, make decisions and take action, so as to improve their circumstances. The ultimate goal is more equitable and sustainable development (Guijt and Shah, 1998/2001, p. 1).

This is a fundamentally different approach than that which EPS encourages. When following EPS, even with the best of intentions, an engineer would

- identify a community's problem or get the problem already defined by someone else such as an NGO (e.g., GIVEN: the community's drinking water is polluted by waste),

- draw technical boundaries and make assumptions about the problem that often exclude its socio-political dimensions (e.g., water problem is about high levels of contaminants in the water, not about poverty or overuse of clean water by a nearby maquila factory),

- exclusively within the technical boundaries of the problem, create a model of the problem and identify scientific principles that apply to such model (e.g., input-output model of water; principles and equations from fluid mechanics and water biochemistry),

- propose a technical solution (e.g., containing wastewater safely) while ignoring political ones (e.g., maquila factory needs to share its clean water with community),

- design the appropriate technology (e.g., composting toilets), and

- deposit the technology in the community (install the toilets and "train" the community how to use and repair them).

- Leave to solve the next problem in another community (leaving intact the structural issues that produced the problem in the first place: poverty and the factory's monopoly over clean water).

As we will see in Chapter 6, if engineers had followed this approach to the project they had in mind at the outset, the project would have resulted in a failure. Instead, by engaging the community and its multiple perspectives, they discovered that the community wanted something completely different. In Chapter 7, we will see how an engineer never drew strictly technical boundaries around water problems but through participatory practices made visible the political and economic dimensions of water usage.

At the same time, we do not expect development engineers to be experts in community engagement approaches so they would do well, as the engineer in Chapter 7 did, to involve social scientists or others with experience in "participatory development" (PD) and "participatory action research" (PAR). Are there opportunities for working with students from other disciplines on your project? How about with social science or humanities faculty? What might you learn about approaches outside of EPS that could help you better serve communities?

Exercise 30 *It may feel daunting to try to prepare to answer these questions on your own. But you're not alone! Consider the three recommendations below as a way of breaking your responsibilities into concrete, achievable steps. Which recommendation could you take action on now? Make a commitment to complete that step over the next week.*

Recommendation 1: *Turn to the end of Chapter 5 and consult the list of references available to you on community participation. Obtain one of these references from your library or the web and read it. Make note of lessons learned or suggestions that might guide your own work.*

Recommendation 2: *Find out if there are students or faculty on your campus who have been involved in development work and who have some background in community participation approaches. Arrange to meet with them and ask them about their experiences, and for their advice. As engineer Donna Riley puts it, "The value of working with social scientists can not be underestimated" (Riley, D., 2007, p. 12).*

Recommendation 3: *Begin an "engineering for community" notebook. Track your concerns about your ESCD work, keep useful articles, and record advice or lessons as you progress here.*

4.6.3   FIGURE OUT WAYS THE TIME-SCALE OF YOUR PROJECT CAN BE EXPANDED

Practically speaking, most engineering development projects take place for individual students over the course of a semester, on average (with some being shorter, some longer). However, projects that have a multi-year commitment in a particular community seem to be more successful than those that don't (e.g., see Silliman, S., 2009).

If you are lucky enough to be involved in a project that has a multi-year commitment, as a student, be sure to review all other student work that has preceded yours. You also have the opportunity to interview faculty members and other stakeholders in the projects about some of the community-oriented issues we have raised in this chapter. And, you have the benefit of learning from past mistakes. In Chapter 8, we document the growth, learning, and realizations of an engineering student as she decided to research the history of a student project by interviewing faculty and students and digging deep into the project's archives.

If your project is not multi-year, or if you are involved in the first year of the project, begin thinking from the outset about what you would tell future groups about your experiences. Keep notes about particular learning experiences, blind spots, or areas for future research. In this way, the project develops a sort of "institutional memory" that can be passed on from group to group, providing some continuity.

4.6.4   MAKE PLANS FOR "FAILURE"

This can be a particularly difficult concept for students to embrace, but we think one of the major areas for learning in ESCD work has to do with honestly engaging with failure. Perhaps your group has no plans to involve community members in developing a particular design; perhaps you are unable to effectively navigate the politics of the community you are working with; perhaps your design is rejected outright by the community you were hoping to "help." Performing thought experiments as you go about the multiple possibilities for failure, and how to address them should they arise, can be a useful way to maintain an attitude of humility, an openness to learning and opportunity, and a way to disengage from EPS when necessary.

4.6.5   DESIGN A LANDING PAD

According to Epprecht, anyone involved in work-study experiences abroad should be provided with a structured period for self-reflection and understanding to develop after exiting the field. Although ESCD projects are not the same as work-study projects at other universities, we believe Epprecht's points may be useful for students in ESCD projects. Engineering students in particular may find their commitments to engineering problem-solving challenged by their development experiences. Confusing discoveries can happen when you study or work abroad; the chances for confusion are even greater when you are only abroad for a few weeks, or have less preparation than desired.

To this end, we recommend that you and your fellow students formally incorporate opportunities and exercises that can help you to make sense of your experiences, write about them formally and informally, and present and discuss those results, in groups and individually, with experienced faculty members and other students—especially ones who have more extensive experiences living and/or working in both the guest and host cultures. This sort of "landing pad" could help soften negative effects or misconceptions that emerged during the field experience. Design faculty might need to rethink the format and requirements of design reports written by students at the end of a course. As these reports almost always focus on what worked, what was built, and results, they do not invite students to reflect on their experiences with ambiguity, failures, fears, and confusion.

4.6.6   DEVELOP MEANINGFUL ASSESSMENTS OF YOURSELF AND YOUR PROJECT

As we noted above, big development projects were notorious for having little or no accountability, especially over the long term. They could fail with very few repercussions for the corporations, organizations, or governments involved. According to Easterly, "…nobody is actually held accountable for making *this* intervention work in *this* place at *this* time" (Easterly, W., 2006, p. 369). This should lead us to the question: How do we assess ESCD projects in both the short and long term?

Exercise 31 *From what we have seen, there is very little long-term assessment of most engineering development projects. How do you think you or your group might be able to do things differently? One approach might be to develop a rubric for assessment with your host community, your fellow students, and other key stakeholders. How would you begin to design such a rubric? Whose input do you need as you design it?*

Remember that it is not enough to be in possession of good intentions. Educators and students could learn from failures as well as successes, and use rigorously researched information to improve future projects. Easterly writes,

> The lessons of [research evaluating development aid programs] is that some equally plausible interventions work and others don't. Aid agencies must be constantly experimenting and searching for interventions that work, verifying what works with scientific evaluation. […] The aid agencies must carefully track the impact of their projects on poor people using the best scientific tools available, and using outside evaluators to avoid the self-interest of project managers (Easterly, W., 2006, p. 375).

In that sense, just as a chameleon adapts its body tints to the surrounding environment, ESCD project workers should also seek to tailor their participatory, community engagement approaches to local contexts and needs. A one-size-fits-all approach lacks the necessary adaptability to function in varied circumstances.

Exercise 32 *Think about an engineering development activity you have been (or will be) involved in or would like to be involved in. Design three exercises that could be integrated into the normal design exercises you must complete for the projects and that will help you to make community and self-reflection more central parts of your projects. These could be assessment rubrics, journal prompts, or failure charts.*

4.7   CONCLUSION

The aim of this chapter was to take into account the realities and challenges that students working on community development projects must face. Those realities include numerous competing requirements and interests: finishing schoolwork on time and with good outcomes; wanting to feel good about a project; and satisfying multiple stakeholders, such as professors, parents, community members, NGOs, and fellow students. What we have tried to argue is that, given these many demands, students, and their professors must find ways to integrate and make central community concerns.

For example, one of the major challenges we have highlighted is the engineer's commitment to EPS as an approach to problem-solving. Although EPS may often be well suited to problems from engineering science textbooks, it may be counterproductive in development contexts if it is not complemented by other approaches, such as participatory development. But exploring participatory development requires that students enter territory in which problems and communication styles may be unfamiliar, open-ended, or slow to emerge. Engineers in these contexts may find themselves not in the position of expert, but in the position of novice, because so many issues—cultural values, language, ways of being—are difficult to know and understand in a short time period.

The challenge for you is to figure out how to balance what is required of you as an engineering problem-solver with other ways of being in the world. How can you incorporate being an excellent listener into your work as an engineer? How do you maintain your flexibility when working with a community whose desires may be different than the "needs" you had originally imagined? How do you maintain your professionalism when things move slowly, or change course, or must be abandoned altogether? How do you make room for getting to know these new people? At what point can you say you know them enough to intervene in their lives? Can you afford not to? Can they?

Perhaps, most importantly, what might these important lessons teach you about being an engineer in general? What new values or approaches have you developed? How have you changed as a person, citizen, or engineer?

These are major questions, questions that get to the heart of what it means to be an outsider in a new community, what it means to be an engineer, and what it means to be a human being with responsibilities to a "global community." In the next chapter, we expand on the ideas introduced here in order to give you some practical approaches for interacting with others in engineering development contexts.

REFERENCES

Amadei, B., and Wallace, W. A. (2009). Engineering for Humanitarian Development. *IEEE Technology and Society Magazine,* **28**(4), 6–15. 93

Baillie, C. (2006). *Engineers within a Local and Global Society*, Morgan & Claypool Publishers. 94

Bridger, J. C., and Luloff, A. E. (1999). Toward an interactional approach to sustainable community development. *Journal of Rural Studies,* **15**(4), 377–387. 85

Burkey, S. (1993). *People First: A Guide to Self-Reliant Participatory Development.* London and New Jersey: Zed Books Ltd. 86

Catalano, G. (2007). *Engineering and the Other America.* Paper presented at the American Society for Engineering Education. 101

Chambers, R. (1997). *Whose Reality Counts?: Putting the First Last.* London: ITDG Publishing. 87, 107

Chhetri, N. (2009). *Lessons from Ghana: Why Some Technological Fixes Work and Others Don't.* CSPO Soapbox. Retrieved September 28, 2009, from the World Wide Web: `http://www.cspo.org/soapbox/view/090827P6KF/lessons-from-ghana-why-some-technological-fixes-work-and-others-dont/` 98

Cooke, B. and Kothari, U. (Eds.) (2001). *Participation: The New Tyranny?* London and New York: Zed Books. 105

Crawley, H. (1998/2001). Living Up to the Empowerment Claim? The Potential of PRA. In I. Guijt and M. K. Shah (Eds.), *The Myth of Community* (pp. 24–34). London: ITDG Publishing. 105

Downey, G., and Lucena J. (2006). "The Engineer as Problem Definer: How Globalization, Underrepresentation, and Leadership are Variations of the Same Problem." Invited Distinguished Lecture, American Society of Engineering Education annual conference. Chicago, IL. 92

Easterly, W. (2006). *The White Man's Burden: Why the West's Efforts to Aid the Rest Have Done so Much Ill and so Little Good.* New York: The Penguin Press. 96, 99, 108, 112

Epprecht, M. (2004). Work Study Abroad Courses in International Development Studies: Some Ethical and Pedagogical Issues. *Canadian Journal of Development Studies,* **XXV** (4), 687–706. 99, 102, 105

Esteva, G. (1992). Development. In W. Sachs (Ed.), *The Development Dictionary: A Guide to Knowledge as Power,* (pp. 6–25). London and New York: Zed Books. 107

Guijt, I. and Shah, M. K. (Eds.) (1998/2001). *The Myth of Community: Gender Issues in Participatory Development.* London: ITDG Publishing. 87, 105, 109

Heron, B. (2007). *Desire for Development: Whiteness, Gender, and the Helping Imperative.* Waterloo, Ontario, Canada: Wilfrid Laurier University Press. 105

Hickey, S. and Mohan, G. (2004). *Participation: From Tyranny to Transformation.* London and New York: Zed Books. 105

Illich, I. (1968/1990). To hell with good intentions. In J. Kendall (Ed.), *Combining service and learning: A resource book for community and public service* (Vol. 1, pp. 314–320). Raleigh, N.C.: National Society for Internships and Experiential Education. 107

Jackson, J. T. (2005). *The Globalizers: Development Workers in Action*. Baltimore, MD: The Johns Hopkins University Press. 99

Maslow, A. (1943). A Theory of Human Motivation. *Psychological Review, 50*(4), 370–396. 99

Mathie, A. and Cunningham, G. (Eds.) (2008). *From Clients to Citizens: Communities Changing the Course of Their Own Development*. Bourton on Dunsmore, Rugby, UK: Intermediate Technology Publications Ltd. 86, 87, 107

McKenzie, S. (2004). *Social Sustainability: Towards Some Definitions*. Magill, South Australia: Hawke Research Institute, University of South Australia. 86

Mulder, K. (2006). Engineering curricula in Sustainable Development. An evaluation of changes at Delft University of Technology. *European Journal of Engineering Education, 31*(2), 133–144. 101

Page, B. (2003). Communities as the agents of commodification: The Kumbo water authority in Northwest Cameroon. *Geoforum, 34*, 483–498. 105

Riley, D. (2007). *Resisting Neoliberalism in Global Development Engineering*. Paper presented at the American Society for Engineering Education. 92, 110

Riley, D. (2008). *Engineering and Social Justice*. San Rafael, CA: Morgan & Claypool. 95, 99

Rist, G. (2002). *The History of Development*. London and New Jersey: Zed Books. 99, 103, 104

Sachs, W. (1992). *The Development Dictionary*: St. Martin's Press. 104

Salmen, L. F. (1987). *Listen to the People: Participant-Observer Evaluation of Development Projects*. New York: Oxford University Press. 86

Salmen, L. F. and Kane, E. (2006). *Bridging Diversity: Participatory Learning for Responsive Development*. Washington, D.C.: The World Bank. 86

Schneider, J., Leydens, J. A., and Lucena, J. (2008). Where is 'Community'?: Engineering education and sustainable community development. *European Journal of Engineering Education, 33*(3), 307–319. 99

Schneider, J., Lucena, J., and Leydens, J. A. (2009). Engineering to Help: The Value of Critique in Engineering Service. *IEEE Technology and Society Magazine, 28*(4), 42–47. 92

Scott, J. (1998). *Seeing Like a State: How Certain Schemes Improve the Human Condition Have Failed*. New Haven. London: Yale University Press. 103

Selingo, J. (2006). May I Help You? *ASEE Prism, 15*(9). 99

Sichel, B. (2006). *'I've come to help': Can tourism and altruism mix?*, Briarpatch: Fighting the War on Error. Retrieved May 21, 2009, from the World Wide Web: `http://briarpatchmagazine.com/2006/11/02/ive-come-to-help-can-tourism-and-altruism-mix/` 99, 108

Silliman, S. E. (2009). Assessing Experiences of International Students in Haiti and Benin. *IEEE Technology and Society Magazine,* **28**(4), 16–24. 111

Vandersteen, J. D. J., Baillie, C. A., and Hall, K. R. (2009). International Humanitarian Engineering: Who Benefits and Who Pays? *IEEE Technology and Society Magazine,* **28**(4), 32–41. 86

Visvanathan, S. (2005). Knowledge, justice, and democracy. In M. Leach, I. Scoones, and B. Wynne (Eds.), *Science and Citizens: Globalization and the Challenge of Engagement* (pp. 83–94). London and New York: Zed Books. 104

CHAPTER 5

Listening to Community[1]

At the first Senior Design team meeting in the fall semester, the enthusiasm was almost palpable. Although several engineering students had requested to work on a community development project, only a few had been chosen. For the project, village leaders in Honduras had requested a system that would bring clean drinking water to their remote village. One of the five team members summed up the spirit of the team when she said, "We've waited four years for this. Finally, we get a chance to *do* something real and meaningful with our engineering knowledge." Eager to begin, the Senior Design team brainstormed several possible types of water systems. Before long, they were discussing whether their spring visit to the village could be an implementation visit.

That discussion soon changed. At their next meeting, the Senior Design instructor listened to the team's goals and said, "We know the village leaders want the water system. But how do we know that this is what the whole community wants? And if so, what kind of system does the community prefer? What are their perspectives?" Silence filled the room. No one could answer those questions. In their enthusiasm, the team realized that they had forgotten about listening to the perspectives of those who would be most impacted by the system.

To their credit, the team quickly rebounded. Their spring visit involved the Spanish-speaking team members conducting face-to-face surveys, including house-to-house visits, and community meetings to better understand more about community members' perspectives.

Another team of students was bound for a village in India, where a civil engineering faculty member led their work on a water and sanitation project. Or so they thought. Once they arrived, they listened to the village in a participatory and consensus-building process designed to identify and define community desires. That process unveiled a completely different project than the one initially proposed by village leaders: once all perspectives had come to the fore, the village actually preferred a power-generating windmill. (We will say more about this case in Chapter 6, including about the complexities associated with the community's choice.)

Both of these are actual community development cases that underscore the importance of the willingness to listen to multiple perspectives. However, their existence should not give the impression that effective listening is commonplace in sustainable community development (SCD) contexts. Indeed, as the previous chapter on community accentuated, the history of SCD is rife with project failures. William Easterly highlights this point when he illuminates the tragedy of a half-century of well-intentioned but often ill-conceived development, which resulted in the West

---

[1] Some parts of this chapter originally appeared in a publication by Leydens and Lucena, "Listening as a Missing Dimension in Engineering Education: Implications for Sustainable Community Development Efforts," in *IEEE Transactions on Professional Communication*, Vol. 52, No. 4, December 2009, pp. 359–375.

spending 2.3 trillion in foreign aid, yet the development industry "still had not managed to get twelve-cent medicines to children to prevent half of all malaria deaths,…[or] to get four-dollar bed nets to poor families,…[or] to get three dollars to each new mother to prevent five million child deaths" (Easterly, W., 2006, p. 4).

Reasons for those failures are complex and multifaceted, and this chapter explores one of them: failure to listen effectively to community perspectives. Unfortunately, examples of engineering for development cases wherein ineffective listening occurred are all too common (e.g., Adas, M., 2007; Easterly, W., 2006; Jackson, J., 2005; Shiva, V., 1993; see also Chapter 4). After briefly exploring two such cases, we will

- Discuss how listening is positioned within engineering education,

- Define and describe the dimensions of contextual listening,

- Identify barriers to and benefits of contextual listening, and

- Propose an alternative problem-solving, listening-centered approach suited to SCD contexts.

Exercise 33 *How would you define effective listening within the context of SCD projects? List what you consider the primary components or dimensions of effective listening.*

5.1   LISTENING IN BIG DEVELOPMENT: THE EL CAJÓN DAM CASE

How do large development organizations, such as the World Bank, listen to the people they are supposed to serve? What's going on when these organizations allow public comment on development projects beyond and within the affected communities?

Consider the case of El Cajón Dam. In his meticulously researched book *The globalizers: Development workers in action,* Jeffrey Jackson tells how major players in the development industry, including the World Bank (WB), Inter-American Development Bank, and corporate dam contractors, carried out the design and build phases of El Cajón. Situated along the Humuya River in Honduras, El Cajón Dam, completed in 1984, was designed to decrease dependency on foreign oil and provide enough electricity not only for Honduras but for Honduras to sell to neighboring countries.

Yet during pre-implementation, the project received much public protest, including from Honduran engineers and policy makers. For instance, organizations such as the Honduran Society of Civil Engineers and the Honduran Forestry Commission expressed concern about the risky nature of the dam project. Even the union of the Honduran national electric utility, which stood to be a primary recipient of the dam's potential wealth-generating capacity, opposed the dam project for being too large.

The WB itself recognized that the project cost was tremendous relative to the size of the Honduran economy; the total project cost would constitute over 50 percent of Honduras' annual economic output and four times its annual government revenues. By contrast, a similar project in the US would cost a much smaller fraction of such output and revenues (Jackson, J., 2005).

Further, the WB was aware of the public opposition to El Cajón. For instance, during the public comment phase, the WB received letters from two Honduran engineers, who echoed local community perspectives in opposing El Cajón on three grounds: "its high cost, potential for endangering the downstream population in the event of a failure of the dam, and above all, diverting scarce resources from other development needs" (Jackson, J., 2005, p. 168 quoting WB internal correspondence). However, for multiple complex reasons, officials at the WB were not interested in listening to diverse perspectives. Regarding the Honduran engineers' concerns, WB staff wrote in internal correspondence that "The Bank's files contain two unsolicited screeds by Honduran engineers.... Parts of these obliquys [sic] are quite poetic..." (Jackson, J., 2005, p. 167). A screed refers both to a personal letter and to a ranting piece of writing. Obloquies refer either to condemning or abusive language or to a situation in which someone is discredited or has a bad reputation. WB staff continued the dismissive, patronizing tone by noting, "Such complaints are not unusual, and the Bank's refusal to be drawn in to a debate was the correct stance" (Jackson, J., 2005, p. 168).

In many ways a dam designer's dream, El Cajón held much promise (see Figure 5.1). Its designers said it would serve not only to bolster the Honduran energy infrastructure but also to enable Honduran self-sufficiency and the capacity to generate revenue from excess electricity reserves. Technologically, the dam was considered state-of-the-art at the time (mid-1980s). At 741 feet tall (nine feet taller than Hoover Dam in the US), the dam was the eighth highest in the world. It "spanned the 1,253-foot wide canyon using an 'elegant doubly curved maximum cantilever' variable radius parabolic arch design 'based on membrane theory'" (Jackson, J., 2005, p. 163). According to one project consultant, the dam represented "a half century's progress in concrete arch dams" (Jackson, J., 2005, p. 163). Further, feasibility studies identified the Humuya River as the site most likely to produce "maximum power-generation potential at the lowest construction cost" (Jackson, J., 2005, p. 156). Flood control and irrigation were also touted as benefits. The estimated $35 million in petroleum savings would mean the cost of the dam would be paid off in roughly 20 years (Jackson, J., 2005, p. 162), as the four turbines combined would produce 300 megawatts of electricity daily.

Despite its great promise, the outcome of El Cajón was disastrous. Although the reasons for its failure are too many to describe here, many of them aligned with the local perspectives. Such perspectives consistently included resource concerns:

- The dam's high cost placed Honduras in tremendous debt, which had an exponential effect when Honduran currency later devalued.

- When the predictions for rising oil prices were not fully realized, expected revenues were smaller.

Figure 5.1: El Cajón Dam in Honduras (Wikimedia Commons, 2009).

- Much of the public protest also centered on placing too many of Honduras' energy eggs in a single basket. Even some engineers working on the El Cajón project were in favor of an alternative plan to build five smaller dams at diverse points around Honduras (Jackson, J., 2005, p. 167).

However, largely for bureaucratic reasons, the WB preferred to make large loans (Jackson, J., 2005, p. 153).

In 1986, just two years after the dam's completion, one of the four turbines failed and was sent to Switzerland for costly repairs; in the same year, cracks appeared in the cement grout used to plug holes in volcanic rock near the dam, and those were fixed by 1990—after Honduras had dished out another $72 million for repairs. Since the desired 50 or 100 years of water level data were not available, water level estimates had been extrapolated from rainfall data from 1967 to 1978 (Jackson, J., 2005, p. 158). By 1996, droughts had left the reservoir two-thirds full, one hundred feet below expected water levels, and the dam was operating at 60 percent of its electricity-generating capacity (Jackson, J., 2005, p. 151). The dam has since been plagued by other problems.

What role did listening play in the failure of El Cajón Dam? Jackson indicates that the WB's response to public opposition

implies the bank had no responsibility whatsoever to reply to the concerns of Honduran citizen's groups opposed to the plan, and the very grounds on which El Cajón was being criticized by these groups did in fact turn out to be valid. Indeed, the same [WB] report later concludes that these 'dissidents' were right (Jackson, J., 2005, p. 168).

This case reinforces the importance of the responsibility of designers, especially SCD designers, to listen and be accountable to the diverse stakeholder perspectives that arise within any SCD project, large or small. Certainly all the problems associated with the El Cajón Dam cannot be blamed on the failure to listen to local perspectives. However, that failure is a significant component in the El Cajón case. History might very well have been different if instead of dismissing legitimate concerns as "screeds" and "obloquies," the WB had seriously listened to those perspectives and included them in the debate.

Unfortunately, the lack of effective listening seen in the El Cajón Dam case is all too common in engineering for development work, as manifested by cases involving different circumstances, times, and contexts—but similar outcomes (e.g., Adas, M., 2007; Burkey, S., 1993; Easterly, W., 2006; Shiva, V., 1993; Slim and Thomson, 1995; Salmen, L., 1987; Mason, K., 2001).

5.2   LISTENING IN LITTLE DEVELOPMENT: BRICK MAKING KILNS IN PESHAWAR, PAKISTAN

Perhaps you are thinking that listening is difficult in the context of large development projects such as El Cajón because of the bureaucratic and political nature of large development loans. Perhaps listening happens more easily in smaller development projects. Yet this might not be the case either. In another instance, technology transfer also had negative effects, and engineers' lack of listening was part of the problem (Mason, K., 2001). In the early 1990s, an attempt was made by well-intentioned community development organizations to modernize the brick making operations in Peshawar, Pakistan, by replacing what was considered an outmoded, inefficient kiln with a vertical shaft brick kiln (VSBK), seen in Figure 5.2. Compared to the "outmoded" kiln, the VSBK had higher energy and fuel efficiency and lower pollution emissions.

Despite that technological promise, the kiln modernization process failed for a number of reasons. First, no one listened to the local perspectives on the scale of their brick making operations: the VSBK had a lower production capacity (4,000-7,000 bricks in 24 hours) than the old kilns (7,000-28,000 bricks in 24 hours), so it was less suitable to the medium-scale operations in Peshawar.

A second reason had to do with training. Fortunately, the consulting Chinese engineers who had perfected the VSBK built and ran the kiln and conducted on-site training. However, the training lasted only a few weeks, which did not allow enough time to work out significant kinks in the production process. Further, the Chinese engineers did not spend enough time listening to the local brick makers' knowledge of why their old system worked well. As it turns out, the quality of the local coal and clay was different enough from Chinese coal and clay that the VSBK in Peshawar produced many over- or under-fired or broken bricks. Clearly, a more thorough and participatory investigation of local practices, including both soil and coal tests as well as listening to local perspectives on current practices, could have produced more favorable results (Mason, K., 2001).

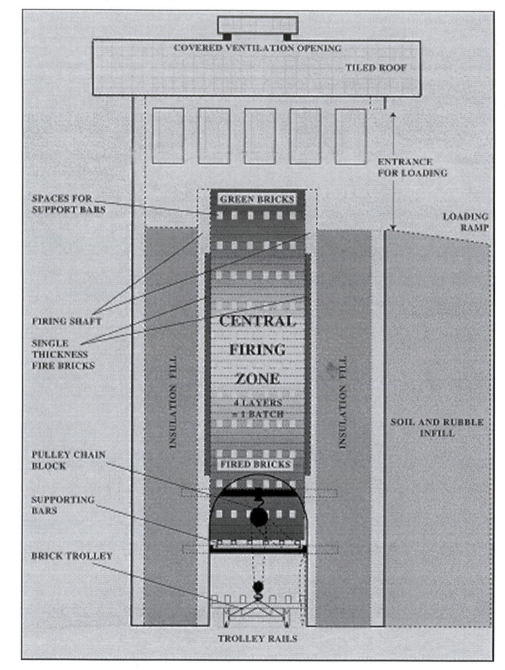

Figure 5.2: Cross-section of a VSBK with single shaft, chain block unloading. (Source: http://www.basin.info/gate/vertical.htm Credit: GATE International.)

In El Cajón, in the brick making case, and in other cases, the project outcomes could have more successfully promoted sustainability, community, and development—if there had been contextual listening to community perspectives. Contextual listening is described in more detail later in this chapter.

5.3   WHERE IS LISTENING IN ENGINEERING EDUCATION?

Before diving into the kind of listening required for effective work in SCD, it is important to understand the place that *listening* occupies in your engineering education. The exercises below serve as an opportunity to explore the role of listening in engineering education curricula.

Exercise 34 *What role has listening instruction played so far in your engineering education? As background, consider these facts:*

- *The National Academy of Engineering's profile of the Engineer of 2020 accentuates listening. Specifically, they "…envision a world where communication is enabled by an ability to listen effectively as well as to communicate through oral, visual, and written mechanisms" (National Academy of Engineering, 2004, p. 55).*

- *Leaders behind Project Kaleidoscope, an initiative designed to transform US undergraduate science and technology education, underscore the importance of empathy as "critical to effective collaboration, building trust and resolving differences in viewpoint. It also requires the cultivation and use of what is probably our most neglected communication skill: listening" (Astin et al., 2003, p. 13).*

- *In the American Society of Civil Engineering (ASCE) body of knowledge for the 21st century, one of the professional outcomes focused on "[m]eans of communication [that] include listening, observing, reading, speaking, writing, and graphics." The ASCE even goes so far as to stipulate that listening and other fundamentals of communication "should be acquired during formal education" (American Society of Civil Engineers., 2004, p. 135).*

- *The Accreditation Board for Engineering and Technology (ABET) requires that engineering programs ensure their graduates are able to "communicate effectively," which implicitly includes listening (ABET, 2004).*

How well have the required and non-required components of your engineering curriculum lived up to these engineering association ideals?

A recent survey of practices in communication programs at seven technical institutions found that listening was not emphasized in any of those programs (Leydens and Schneider, 2009). Generally, attempts to teach engineering students to listen are often located in senior design courses.

Exercise 35 *In many senior design courses, listening is conceptualized as hearing or paying attention to the customer or the client (i.e., as a basic skill). Think about how your experience in design courses compares with the results of one engineering education study, which reports that,*

skills in listening (incoming) and speaking (outgoing) are strongly linked, yet listening in-
struction and practice is scarcely to be found in the curriculum. Nowhere is this disconnect more
clearly represented than in the Senior Design teams. Generally, everyone wants to present his
[sic] point of view—and can do so forcefully—but has great difficulty in listening to and accept-
ing the ideas of others. Communication falters among team members as a result (Wikoff et al.,
2004).

If listening is *important* for a twenty-first century engineer, it is *essential* for engineers involved
in SCD contexts. This conclusion is evidenced by the four cases mentioned above—an overly eager
but ultimately refocused Senior Design team, the team that engaged and listened to the Indian
community to learn that the community actually desired a power-generating windmill, and the El
Cajón Dam and brick making cases. One primary component of avoiding top-down SCD projects
that fail to make culturally responsive inquiries into community perspectives involves the ability to
listen *in* and *to* context, or contextual listening.

5.4   WHAT IS CONTEXTUAL LISTENING?

If listening is critical for engineers, what exactly is listening? In this section, we contrast two types
of listening, basic and contextual listening. Basic listening is *necessary* in any human communicative
interaction but is not *sufficient* for listening in SCD (and arguably, most other) contexts.

Key Term

Basic Listening refers to hearing or paying attention to the verbal and nonverbal messages of any
speaker, such as a client, customer, local community member, coworker, or instructor. Basic listening
is framed as a dyadic process of speaking (output) and hearing/receiving information (input). In this
form of listening, relevant information is generally reduced to specific and quantifiable requirements
such as cost, weight, technical specs, desirable functions, and timeline. Contextual and qualitative
information, such as the history or political agenda of the person(s) making the requirements, is
often devalued or ignored altogether.

A strong connection exists between listening to and engaging with local community members' perspectives and desires (or failing to do so) and the degree of ownership, success, and long-
term sustainability of community development projects (e.g., see Salmen, L., 1987; Burkey, S., 1993;
Slim and Thomson, 1995; Salmen and Kane, 2006). As we noted at the outset of this chapter, case
studies point out the pitfalls of not listening to local perspectives or not acting on those understandings (see also Ogundimu, F., 1994; Starosta, W., 1994).

Key Term

Contextual Listening: A multidimensional, integrated understanding of the listening process wherein listening facilitates meaning making, enhances human potential, and helps foster community-supported change. In this form of listening, information such as cost, weight, technical specs, desirable functions, and timeline acquires meaning *only when* the context of the person(s) making the requirements (their history, political agendas, desires, forms of knowledge, etc.) is fully understood.

The absence of basic or contextual listening is satirized in Figure 5.3.

"We development experts have arrived,
and we know what you need:
one good piece of earth-moving equipment!"
© David Williams, used with permission.

Figure 5.3: Effective communication requires listening.

Below we describe several dimensions (both characteristics and desired outcomes) of contextual listening. These dimensions are interrelated, overlapping, and sometimes interdependent.

Exercise 36 *Search for visual representations of listening in engineering textbooks, websites, and other materials. To what extent do these representations align more with a basic or a contextual listening model? To what extent is your understanding of listening shaped by these models?*

Since each SCD project is unique, for each dimension, we have included salient questions in the Appendix that, if explored thoughtfully, can foster contextual listening across multiple contexts. Sometimes such questions are partially addressed through pre-travel research, other times via on site interviews and discussions, and frequently from spending time—and being with—people on site. Overall, the questions help us better focus on what issues in SCD projects are most crucial to listen for and to, and we recommend that SCD practitioners ask themselves these questions *throughout* their SCD projects.

Definitions of SCD center on "the importance of striking a balance between environmental concerns and development objectives while simultaneously enhancing local social relationships" (Bridger and Luloff, 1999, p. 381). The dimensions of contextual listening below provide some ideas for moving toward just such a balance.

Exercise 37 *Think of one time in your life when effective listening made a significant difference in a project, relationship, or situation. Also, think of a time when the absence of effective listening negatively affected the outcome of a project, relationship, or situation. From these two events, what can you learn about the dimensions of listening?*

5.5   DIMENSIONS OF CONTEXTUAL LISTENING

In past years, students and faculty in our ESCD course (described in Chapter 8) have asked three types of questions related to listening:

- In SCD contexts and especially while on-site, what are you listening for—to understand what, exactly?

- What are some of the characteristics of an effective [contextual] listener?

- What are the desired outcomes of [contextual] listening?

Collectively, the dimensions below are designed to address these three questions.

A. Integrating History and Culture

No SCD engineering project occurs in a vacuum. Instead, such work occurs in a community context, which itself is shaped by international, national, regional, and local socio-cultural/historical contexts. One can imagine these surrounding contexts as concentric circles, with the local community at the center, as in Figure 5.4. Such contexts materialize out of lived experiences and interpretations of such experiences and tend to vary significantly in the degree to which they influence local perceptions and identities. Knowing as much as possible about the history and socio-cultural realities of people in and around the community where your SCD project occurs is vital to project success—and is frequently overlooked. Since interpretations of history and culture vary based on one's perspective, it is important to both consult background readings and listen to how community members frame understandings of their own history, culture, and current perspective (for one method of doing so, see PDS at the end of this chapter).

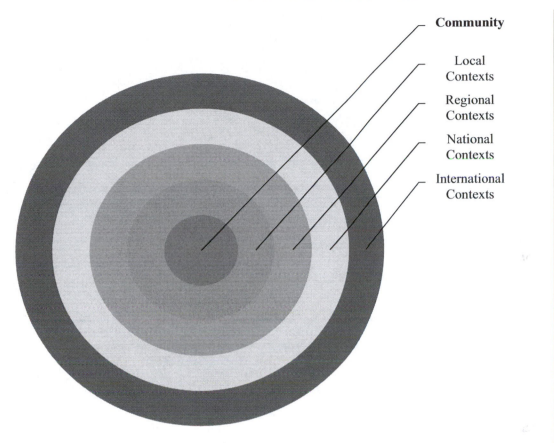

Figure 5.4: A simplified community-centric model of SCD.

Certain questions accentuate the importance of listening in and to context. Such questions are listed in full in the Appendix and a few appear after each dimension below, along with an opportunity to apply ideas regarding each dimension.

Questions for Integrating History and Culture

- What are the origins of the community that you are hoping to serve? What are the different ways to educate oneself to search and listen for these origins?

- What is the history of your relationship to the community? How did your project get to be there in the first place? Was it invited by the community? Was it proposed by your faculty or church?

- As you listen, how do community members indicate whether issues of gender, culture, nationality, social class, and race/ethnicity inform the community's diverse outlooks on itself, SCD, and outsiders?

Application

- How might asking the above questions have made a difference in the El Cajón dam and Peshawar brick making cases?

B. Being Open to Cultural Difference and Ambiguity

Acquiring traits of an effective listener is a worthy ideal. However, even with those traits, one may still be unable to achieve effective listening. Listening is more than traits; it is also an attitude or state of being that involves openness to the kind of ambiguity borne of cultural contrasts. The questions below encourage reflection on such a state, largely achieved through self-reflection and experience.

Questions on Being Open to Cultural Difference and Ambiguity

- What are your strengths and limitations as a listener?

- How can you constantly reassess your own degree of openness to perspectives that differ from your own?

- How tolerant are you about ambiguity, i.e., about not seeing some aspects of the world in absolutes, as either right or wrong?

- How do you deal with cultural difference? When a student from another country highlights differences between his or her culture and yours (in food, customs, values, etc.), do you try to make those *differences* into *similarities*? Or do you accept the differences as they are, even at the cost of some discomfort and ambiguity?

Application

- In the El Cajón case, to what extent was the dismissal by World Bank technocrats of the concerns by local engineers an act of intolerance to differences in technical assessment?

- In the Peshawar brick making case, to what extent was the reluctance by Chinese engineers to more thoroughly explore local brick making practices an unwillingness to understand and perhaps value local knowledge and cultural differences?

C. Building Relationships

In contextual listening, an emphasis exists on forming effective interpersonal relationships built on trust. Such relationships enable people to work successfully toward mutual goals. For instance, participant researchers of an educational intervention in post-tsunami Banda Aceh, Indonesia, note their interest in

> understanding listening as a practice that extends beyond simply hearing words. Our use of **listening** suggests that [SCD workers] attend to individuals, the classroom as a group, the broader social contexts, and to silence and acts of silencing…. Listening is fundamentally about being in relationship to another and through this relationship supporting change or transformation (Schultz and Smulyan, 2007, p. 100).

Gustavo Esteva, a community organizer and intellectual in Oaxaca, Mexico, underscores another way to think of building relationships that enable transformation. In 2008, when he spoke with well-intentioned students in our ESCD course, he said, "Don't come here [to Oaxaca] to help! Come here to listen, to find out if *our* struggles are *your* struggles. Then and only then, we can sit and discuss how, if at all, we can work together." A clear understanding of shared struggles takes time—and trust—before it emerges. In fact, as the questions in the Appendix suggest, significant obstacles can exist to identifying shared struggles.

As a relational and transformation-oriented act, contextual listening changes *you* and *your* relationship with community. You begin to change because, even if briefly, you begin to see the issue at hand as the community sees it. From this new perspective, you can rearticulate what you have heard back to the community. When you do, community members can now see that you have truly listened and perhaps have come to understand their struggles, even if your struggles are different. If the community confirms that your understandings are accurate, they will be more likely to begin to trust you and your relationship with them will be transformed. You may have reached a dynamic consensus that can be applied in partnership with the community. Such listening is crucial because if local community members see no tangible benefits to sharing their perspectives, they may stop talking or simply tell you what you want to hear; they may mistrust you. Their actual perspectives are then lost, perhaps while also losing an empowered sense of self-determination, which will certainly affect the success of any SCD project (Slim and Thomson, 1995).

Questions on Building Relationships

- How do you develop and maintain trust, the glue of effective relationships?
- How willing are you to change (e.g., become more empathic) to build more trusting relationships with others?

- How comfortable are you establishing positive personal and working relationships with people from other cultures?

Application

- In the El Cajón case and Peshawar brick making cases, how could have the World Bank technocrats or the Chinese engineers established trust with the communities that they were supposed to serve? What do such actions have to do with listening?

D. Minimizing Deficiencies and Recognizing Capacities

When we listen, incoming words and ideas are filtered through our own frames of reference. Two large frames are contrasted here, the *deficit* and *capacity* models. SCD projects operating under a deficit model conceptualize the local community members primarily in terms of what they lack, which can mask the community's capacities. By contrast, the capacity model acknowledges constraints but remains focused on the human, technical, cultural, and other capabilities available to achieve community-driven objectives. In this model, all participants remain open to discovering new or previously unrecognized capabilities that facilitate project success. Contextual listening in SCD contexts requires us to shift our focus from what communities lack to also include what they have. By recognizing a community's history and culture, valuing cultural differences, and building trusting relationships with them, you will begin to see more and more value in the resources that the community has to offer, especially non-financial assets like resourcefulness, techniques for doing things, different forms of organizing and managing time, resources and labor, and insights on how to work *with* nature.

Questions on Minimizing Deficiencies and Recognizing Capacities

- What kinds of listening reflect a deficit model? What kinds reflect a capacity model?

- Might living in a consumer-driven economy, where most comforts are taken for granted and often become "necessities," lead us to see some developing communities as lacking?

- Might our emphasis on valuing assets exclusively in financial terms (How much will it cost? How much will it return?) lead us to undervalue non-financial assets?

- If non-local SCD practitioners genuinely seek to enter the cultural frames of reference of local community members, what local knowledge, practices, and resources emerge as capacities?

Application

- Reflect on how these two models can make your own learning experiences very different. When professors teach from the deficiency model, none of what you bring to the classroom counts. Your previous experiences with the subject, your family history, your informal knowledge—all these things are considered irrelevant. Under the deficiency model, students are viewed as empty vessels to be filled by the expert knowledge of the professor and/or textbook.

- Meanwhile, how is teaching and learning different when professors teach from the capacity model?

E. Foregrounding Self-Determination

Community self-determination can be compromised or enhanced by various forms of listening. **Self-determination** suggests that the local community has a significant hand in determining its own destiny, free from undue external pressures. If the project truly *comes from* and is led *by and for* the community with the *invited* assistance of others, it has a better chance of fostering local self-determination. Clearly, this requires a different kind of listening. Through contextual listening, you will be able to understand how and by whom in the community a project can be initiated and led, and, perhaps more importantly, when and how you are being invited to participate.

Key Term

Self-Determination: The ability to play a significant role in determining one's own destiny, free from undue or excessive external pressures.

A Māori ceremonial gathering called a *hui* illustrates the principle of self-determination (Bishop, R., 2005). To cover the costs of running the *hui*, ceremony participants make contributions, traditionally, in the form of food, but today more commonly in the form of money. Pivotal to this act of giving in the ceremony is that the gift

> is placed in a position, such as laying it on the ground between the two groups coming together, so as to be able to be considered by the hosts. It is not often given into the hands of the hosts...[and] the process of 'laying down' is a very powerful recognition of the right of others to self-determination, that is, to choose whether to pick it up or not (Bishop, R., 2005, p. 122).

Key Term

Ownership: The significance or meaning attached to community members' sense that they own a given project—that is, that *they* are largely driving a project toward *their* objectives via decision making and action, with *invited* assistance.

Self-determination includes several components, one of which we describe here: ownership throughout the project. **Ownership** refers to the sense of who owns the project, from start to finish. If the community contributes significantly to defining and articulating their understanding of the problem and to brainstorming possible solutions, they have initial ownership. Ownership also comes from the community controlling or having significant input into major project decisions. If the community completes the project thinking that they did most of the work themselves and that their input shaped consensus-building processes and project outcomes, they have likely had project ownership throughout the process. In such a circumstance, the local community is more likely to assume responsibility for maintaining and/or upgrading project-related technologies—which is essential for the project to be *sustainable*. Generally, once project ownership is transferred outside local control, self-determination begins to deteriorate, in *any* phase of the project.

The absence of contextual listening can jeopardize self-determination in several ways. For instance, local community members who do not see their perspectives heard and incorporated into the various project phases (problem definition and solution, implementation, etc.) are more likely to lose a sense of project ownership. Can you think of other ways?

Questions for Foregrounding Self-Determination

- How can you "lay down" your potential contributions to an SCD project without forcing the local community to take them?

- What forms of listening detract from or contribute to self-determination and ownership?

Application

- As a student, how free do you feel to suggest to a professor an alternative assignment that might meet course goals as well as your own learning goals? Typically, commonplace in an independent study, such a suggestion may be less appropriate in core courses but more so in upper-division courses.

- Have you or anyone you know ever made such a suggestion? If so, with what result? If not, what factors may socialize students to not make such suggestions? How does such socialization affect how people listen in situations involving authorities or experts?

- How is listening to people in *any* community connected to promoting community self-determination?

F. Achieving Shared Accountability: How the "ours" vs. "theirs" Becomes OURS

As the previous discussion implies, the sharing of knowledge should not be unidirectional (from non-local to local SCD practitioner) but bi- or multi-directional (see also Ramaswami et al., 2007). Through listening and other means, each group should be learning from each other. This dialogic nature of knowledge exchange fosters power-sharing and shared accountability. That is, if all stakeholders engage meaningfully with the project, they will all feel accountable for their actions and for project outcomes—as the project ideally is a shared mission.

If accountability shifts entirely to the local community, non-local SCD practitioners may risk reifying the history of development, which is fraught with failed development projects wherein development workers had no extrinsic incentives to ensure long-term project success. They leave assuming that someone else will take care of the project. And if accountability rests entirely with the non-local SCD practitioners, the local community has likely checked out of the project—robbing it of the chance to solidify its sense of self-determination, augment its capacities, ensure ownership, and reap long-term project benefits. Herein lies an important paradox of listening in SCD contexts: at the same time one needs to listen so as to place emphasis on *community* self-determination and *local* ownership and benefits, accountability for project outcomes needs to be negotiated and shared by *all* project participants. Exploring and understanding the nuances of this paradox should help team members listen and communicate more effectively. We hope you will strive to know when the project is best located in *their* hands and when it should be placed in *ours*.

Questions for Achieving Shared Accountability

- Who is accountable for the project's success?

- How does this accountability shift over time?

Application

- As a student, who is accountable for the success of your courses? To what degree have you come to accept responsibility for the outcome of a course? What responsibilities are unique to professors, to students, and which are shared?

- What listening and other communicative actions and practices have fostered shared accountability so far in your SCD project? What future actions and practices might you consider?

Two other dimensions of contextual listening, bias awareness and multiple perspective integration, are explained later in this chapter, and questions pertaining to these dimensions appear in the Appendix.

Collectively, the dimensions above and associated questions are designed to help us establish a critical reflexivity with listening, the ability to reflect on how one's listening interfaces with one's status, actions and decisions throughout the project. As you may have noticed, underlying each dimension is a way of conceptualizing people that is designed to collapse and transcend the dichotomy between "insiders" and "outsiders" (i.e., between local and non-local SCD practitioners). As an ideal, we should be aiming to see each other as collaborators with a shared mission, "people to whom we are bonded through ties of reciprocity" (Narayan, K., 1993, p. 672). Sometimes called a "participatory mode of consciousness" this ideal "is characterized by an absence of the need to separate, distance and to insert pre-determined thought patterns, methods and formulas between self and other" (Heshusius, L., 1996, p. 627). Contextual listening aims toward this worthy but difficult-to-achieve ideal.

Exercise 38

- *If you are currently involved in a SCD-related project, notice how your teammates, faculty members, and others describe the local community. In what ways do their descriptions depict community members? As collaborators with a shared mission? As separate, strange, or alien? As different yet still connected "through ties of reciprocity?"*

- *In the next few weeks, notice how your classmates, faculty members, and others listen to each other. What kind of listening might they be enacting? What could they do differently to move towards contextual listening? What could you do differently?*

- *What factors in an engineers' education might diminish and/or enhance the perceived need to listen to community perspectives in SCD contexts?*

5.6 BARRIERS TO CONTEXTUAL LISTENING

If listening to community member perspectives is crucial to SCD project success, what barriers exist to putting contextual listening into action? Below, we discuss two specific barriers created by the engineering curriculum, drawing from a study that elaborated on these barriers (Leydens and Lucena, 2009). The first curricular barrier is the dominance of closed-ended engineering problem solving (EPS), and the second involves the quality and marginality of open-ended engineering design experiences.

5.6.1 ENGINEERING PROBLEM SOLVING

Engineering curricula rely heavily on math-based quantitative problem solving. But does such problem solving facilitate certain habits of mind, ways of knowing, and methods of inquiry while unintentionally marginalizing others? Some engineering students interviewed in our study suggest that this is a distinct possibility (Leydens and Lucena, 2009). In the ESCD seminar (discussed fully in Chapter 8), students discussed how EPS intersects with one's ability to meaningfully engage with a community. EPS is the dominant six-step engineering method (Given, Find, Diagram, Make Assumptions, Equations, Solve) at the core of engineering curricula and reinforced and valued in engineering textbook problems and exams found especially in engineering science courses (e.g., Hagen, K., 2008). Despite years of engineering education reform, students are still largely graded, rewarded and penalized relative to their mastery of the EPS method (Downey and Lucena, 2006).

In a study interview, Michelle, a junior in chemical engineering, indicated that regularly practicing quantitative problem solving shaped her ways of thinking. She said that EPS "gets drilled into you, [through] the process of repetitive problem solving." Similarly, Jake, a senior in mechanical engineering, also indicated that EPS informs the way he solves problems, in and even outside of engineering contexts. Jonathan, a graduate student in engineering systems, confirmed Michelle's assessment during an ESCD class when he admitted that his job as a statics teaching assistant was "to drill the [EPS] method into [undergraduates'] heads." During an ESCD classroom exercise, engineering students calculated that during their undergraduate years they solve anywhere between 2,000 to 3,000 problems using EPS, depending on their major.

Related research suggests that students' ability to listen, understand, and value perspectives other than their own might be hampered by the preponderance of EPS in the curriculum (Downey and Lucena, 2006). EPS includes no explicit mechanism for listening, other than the initial step of listening to a problem statement in order to figure out the relevant information needed to solve the problem. The history, culture, and identity of the person stating the problem have no relevance. EPS also explicitly creates a boundary between the "technical" and "non-technical" dimensions of a problem, reinforcing the myth that the world of problems can be divided as such and marginalizing "nontechnical" dimensions as less important or irrelevant. In EPS, contextual listening is one such marginalized dimension.

We do not question the need for EPS in the engineering curriculum. However, EPS' *preponderance* in engineering curricula sends the message to engineering students that engineers can draw artificial boundaries around problems. That message can squelch other perspectives, stakeholder voices, and issues that could completely redefine and re-conceptualize the problem at hand—and thus the solution (Downey and Lucena, 2006). In fact, when students encounter open-ended problems in design courses, they often devalue design methods because those do not conform to how students have come to see engineering—that is, as EPS, learned in engineering sciences courses (Downey and Lucena, 2003).

Exercise 39

- *When solving problems in your engineering science courses, how often are you encouraged to*

  - *consider the history and culture of the people behind the problem?*
  - *develop listening traits? develop a listening state of being?*
  - *build interpersonal relationships to establish trust?*

- *How do your responses to the above questions shift if focused on problem solving that is closed-ended (generally one answer) vs. open-ended (multiple viable answers)? How does your response differ if the context is not engineering science courses but internship or cooperative experiences you may have had?*

5.6.2   ENGINEERING DESIGN

Yet if students are to learn any kind of listening in their engineering curriculum, it is most likely to occur not within engineering science but within design courses, which typically feature open-ended problems. Research on listening in engineering design contexts stresses active, participatory listening (Reid and Reed, 2005). However, the design experiences some students in our study recounted suggest that certain qualities in design instruction may also inhibit—or at least not foster—listening abilities (Leydens and Lucena, 2009).

For instance, Lisa, a senior in mechanical engineering, would have preferred the social impacts to be meaningfully integrated throughout her year-long Senior Design course, rather than being worth only about 5% of the course grade and tacked on at the end as an afterthought. Better integration, she said, would have helped her listen to and account for community perspectives throughout.

Jonathan suggested that his Senior Design experience was more about "listening to the spec" (an important skill in any design task) than about contextual listening to clients, teammates, or other stakeholders. By "listening to the spec," Jonathan learned to listen to design specifications (mostly in the form of numerical parameters such as cost) but not to the humans who may have conceptualized or interpreted those specifications differently. The above and other evidence suggests that EPS and some design courses may actually serve as barriers to learning and valuing contextual listening (Leydens and Lucena, 2009).

Exercise 40

- *Given what you have read so far in this chapter and in light of your own experience, what factors in your own engineering education have diminished or enhanced the perceived need to listen to community perspectives in SCD contexts?*

- *Which dimensions of contextual listening have been addressed by your design courses?*

- *What recommendations, if any, might you make for curricular reform to enhance students' contextual listening?*

5.7   BENEFITS OF CONTEXTUAL LISTENING

Although the absence of basic and contextual listening in the engineering curriculum can have negative effects, a richer understanding of contextual listening benefits SCD practitioners in multiple ways: here we discuss how contextual listening 1) counters biases, 2) fosters a community-centric approach to problem defining and solving, and 3) integrates multiple perspectives and sectors.

5.7.1   CONTEXTUAL LISTENING COUNTERS BIASES

Contextual listening can be a useful method of addressing biases (Leydens and Lucena, 2009). After taking an ESCD course, engineering students Michelle, Dave, Jake, and Jonathan were convinced that the listening abilities they learned would make them much more likely to apply them in future SCD contexts and beyond. For instance, Jonathan said that the ESCD seminar transformed his approach to solving problems, even beyond SCD contexts. He said, "I'm [now] more willing to understand those biases [that can occur in SCD contexts], [...including] my biases...and some of the biases that I didn't necessarily understand were biases." In this section, we describe a few biases that come to light via contextual listening.

1) The Documentary Bias

This bias may surface when we ask *what kinds of voices count* in listening. The **documentary bias** is a bias toward the written word, which can marginalize the spoken word (Slim and Thomson, 1995). This bias is especially strong in development work in which entrenched bureaucratic practices require extensive written documentation. Yet in cultures with rich oral traditions, the spoken word often has a strong value. Often in SCD contexts, "[local p]eople are not consulted enough because the main debates take place in documents which they do not write, or in meetings which they do not attend" (Slim and Thomson, 1995, p. 4).

The documentary bias can also blind SCD practitioners to the existence of *tacit knowledge*, an important kind of local *practical knowledge* that cannot be easily documented. Such knowledge depends on knowing how and when to apply certain rules of thumb to particular circumstances in an ever-changing environment. Have you tried documenting how to ride a bicycle? How about trying to explain in writing how to adjust your weight, pedaling, grip and balance when you climb vs. descend on your bike? If this proves difficult for you, who probably grew up expecting much knowledge to be documented, imagine requiring people from strong oral traditions to document their techniques or strategies for growing crops, building shelters, or cooking. In that sense, contextual listening can counter this bias if it allows local community members' voices to be accurately and meaningfully represented in community development documents and/or meetings, even if their experiences cannot be easily captured in writing (Slim and Thomson, 1995). If it is culturally acceptable, sometimes video taping local community members can be an alternative way to document their perspectives and knowledge, especially if those perspectives and knowledge need to be shared with some SCD practitioners unable to travel to the local community. However, video taping runs certain risks (e.g., it may diminish interpersonal trust, may encourage performance over substance, etc.).

2) The Dominant Voices Bias

This and the next bias emerge from asking, *whose voices count* in listening? The **dominant voices bias** acknowledges that "[t]he collective voice of any community tends towards generalizations, simplifications, or half-truths and is dominated by the loudest voices. Like the official document, the community view will tend to concentrate on the concerns of the wealthy, the political elite, and social and religious leaders" (Slim and Thomson, 1995, p. 5).

In most community development projects that we have encountered, students often interact via email with someone who claims to represent the community. But who might that person be if he or she has access to computer communications and has knowledge of written English? It is likely that is a privileged member of the community. How might this person be representing the voices of the less privileged members of the community? Contextual listening can counter this bias because it expressly includes multiple voices, which can be "touchstones against which to review the collective version" and in doing so bring in "a much more subtle appreciation of the divisions and alliances within societies" (Slim and Thomson, 1995, p. 5).

3) The Hidden Voices Bias

Inverting the dominant voices bias gives what can be termed the **hidden voices bias.** Potentially hidden voices can include "the elderly, women, ethnic minorities, the disabled, and children" (Slim and Thomson, 1995, p. 5). This bias occurs whenever the perspectives of such people are excluded from what non-local SCD practitioners claim to be a community consensus viewpoint. To keep such a bias in check, SCD practitioners need to ensure that marginalized people are "no longer hidden voices, but ones that shape the collective community voice" (Slim and Thomson, 1995, p. 5). Experienced SCD workers note that "[s]ometimes the hidden voices are the most important of all" (Slim and Thomson, 1995, p. 5). This importance may stem from their unique perspectives on their own and other community members' needs. Thus, the inclusiveness inherent in contextual listening can help make marginalized voices more central. "At community level, therefore, the testimony of individual voices reveals the experience of hidden groups, and counters the bias of those who speak for or ignore them" (Slim and Thomson, 1995, p. 7). In Chapter 6, we describe the Sika Dhari case study, which shows how dominant and hidden voices biases can impact the outcomes of a project, sometimes without the participating engineers' knowledge.

Since contextual listening instruction can reveal a variety of biases that may arise from power differences (see also Shiva, V. Ed., 1994), development research stresses listening that occurs in the speaker's mother tongue with respect for that person's and culture's ways of communicating (Slim and Thomson, 1995).

Exercise 41

- *What biases have you recognized in yourself, your teammates, or others involved in your current SCD project?*

- *How might contextual listening counter such biases?*

5.7.2 CONTEXTUAL LISTENING FOSTERS A COMMUNITY-CENTRIC APPROACH TO PROBLEM SOLVING

Evidence suggests that contextual listening instruction fostered understanding of a community-centric approach to problem solving. For instance, Dave, a junior in environmental engineering, said the ESCD seminar helped him realize that in SCD contexts,

> the role of listening is crucial because community is the driving force, period. That's it. I mean development shouldn't occur for development's sake. Sustainability is only a question of whether the community can sustain it, so it's the community that gets all the weight. So [if] there's going to be a project, you have to understand the community's needs.

Jake also accentuated listening to community perspectives, but added that such listening is also important in traditional engineering contexts, such as in the floodplain construction in his hometown. In that case, he indicated that public works officials actively

> used the peoples' voice in determining where to put the floodplain. As long as the start point and the end point were relatively the same, the path didn't necessarily matter. So by listening to the community, they were able to plan the path of the floodplain to where it would minimize the amount of either destruction of [personal] property or…the community's public properties…and other recreational centers.

In this sense, contextual listening instruction positioned community at the center of SCD projects. These insights by students have been confirmed by scholars and organizations working at the nexus between sustainability and community development (e.g., Burkey, S., 1993; Slim and Thomson, 1995).

5.7.3 CONTEXTUAL LISTENING INTEGRATES MULTIPLE HUMAN AND SECTORAL PERSPECTIVES

Since failed development projects are associated with mechanistic compartmentalization of community perspectives and foci on a single sector (such as the technical or economic sector alone), community development scholars recommend a more contextual, holistic integration (Salmen, L., 1987; Burkey, S., 1993; Slim and Thomson, 1995; Salmen and Kane, 2006). Slim and Thomson note that

> economic factors do not exist in a vacuum. Social relationships reflect and influence economic and political ones, and an improved understanding of the former can shed light on the latter. The various forms of oral testimony give people the chance to voice their experience of family and work relationships, of friendship, love, sexuality, childbirth, parenting and leisure, culture and religion. These aspects of life, which are central to anyone's understanding of his or her world, are often overlooked in [community development] project feasibility studies, which tend to take a mechanistic view of communities, their

needs and possible solutions. Yet people are more likely to take part in something they value and believe in, and are more willing to invest their time and resources in what is feasible within their current social obligations (Slim and Thomson, 1995, p. 7).

Similarly, some interviewees highlighted the importance of multi-sector or multi-disciplinary approaches. For instance, Jake indicated that "the engineer needs to be more involved on a larger scale [in projects] to be able to integrate more of the information, redefine the problem, [and] have an open sense of what the solutions might be, instead of kind of having a [narrow] tool box."

SCD projects often fail for many reasons; one reason stems from the lack of time, desire, or understanding that development workers have to listen to and wade through multiple, complex human perspectives. Admittedly, a more mechanistic view of development involves far fewer complexities and takes less time. In short, a mechanistic approach is more *efficient*, but it also has a weak historical track record of *effectiveness* (e.g., see Servaes, J., 1991; Jackson, J., 2005; Easterly, W., 2006). Further, not accounting for the richness of human perspectives can lead development workers to make unsituated and decontextualized decisions. Clearly, contextual listening requires someone open to ambiguity and complexity.

Exercise 42

- *Reconsider how you defined effective listening in the exercise at the outset of this chapter. In what ways is your current definition different?*

- *Look at Kelvin Mason's short case study, "A Brickmaker's Experience of Partnership" (Mason, K., 2001, Box 4.2, pp. 39–41). Identify the junctures in this SCD case in which contextual listening was not put to use but could have been. Also, identify which dimensions of contextual listening, if they had been effectively applied, would have likely led to a more successful project outcome.*

5.8   PROBLEM, DEFINITION, AND SOLUTION

We are now faced with a dilemma. If significant limitations exist in EPS and engineering design as problem-solving tools in SCD contexts, what additional and alternative problem-solving approaches might be available that incorporate at least some dimensions of contextual listening? This section addresses that dilemma by presenting a concept called Problem, Definition, and Solution (PDS), designed to illuminate location, knowledge, and desires.

Key Terms

Location refers to one's social location or position, including but not limited to issues of wealth, power, status, gender, family and ethnic background.

Knowledge refers to the types of knowledge diverse stakeholders bring to a shared objective, such as formal and informal, expert and non-expert, scientific and non-scientific.

Desires refer to the type of yearnings stakeholders have, including but not limited to selfish and altruistic desires, career goals, political agendas, and visions for the future.

Engineering studies professor Gary Downey began to develop the PDS approach from his training in cultural anthropology and his commitment to ethnography as a way of listening and via research on technical controversies (Downey, G., 1986, 1988a,b). Analyzing technical controversies presented the challenge of positioning different perspectives without giving epistemic priority to one over another. Within PDS, one remains committed to listening to each perspective regardless of its social **location**/positioning (e.g., rich and poor, powerful and powerless), the kind of **knowledge** that they possess (e.g., formal and informal, expert and non-expert), and the **desires** that they might have (e.g., selfish and altruistic). Downey incorporated PDS as a teaching strategy for students in Engineering Cultures, a course that he co-developed with Juan Lucena (Downey et al., 2006). PDS was designed to help engineers look beyond EPS by mapping perspectives and valuing those that are different (Downey, G., 2008).

More specifically, PDS as an approach to listening includes the following actions:

1. Mapping perspectives by identifying three key elements for each stakeholder: location, knowledge, and desires.

2. Analysis and assessment of the implications of proposed solutions for each perspective.

3. Mediation and perhaps reconciliation of contrasting definitions of problems and solutions for the perspectives involved.

4. Considering how shifting one's perspective might contribute to achieving a solution acceptable to all.

In brief, understanding **location** in SCD contexts involves examining the cultural, historical, ideological, family, and personal dimensions that bring people to be involved in a community development project. Practitioners in SCD contexts should also focus on what **knowledges** (e.g., formal vs. informal, experiential vs. analytical, written vs. oral, tacit vs. codified) all stakeholders bring to community development projects. Non-local practitioners can also learn to assess the strengths and limitations of their own engineering knowledge while realizing the value of diverse local knowledges in communities. Engineers will find that there are many approaches to technical problem solving besides EPS. Finally, SCD practitioners should listen to the **desires** of stakeholders, including their own. One goal involves assessing the degree of overlap between a community's struggles, yearnings, and dreams and those of engineers on SCD projects.

During the ESCD seminar, we challenged students to apply PDS to community development projects. We also recognized the limitations of PDS as a listening approach, such as the fact that PDS has not been tested in SCD fieldwork. Hence, we sought and found a few case studies of engineers working in SCD contexts and analyzed their strategies for listening to community (Schneider and Lucena, 2008a,b). These two case studies constitute Chapters 6 and 7.

Exercise 43

- *If your team is slated to work with an actual community, explore how you interact with and engage that community:*

  a) *What areas of expertise or knowledge might the community have that could be revealed via contextual listening? What information about location and desire could emerge from such listening?*

  b) *How does your own engineering expertise affect how you see the community members? Their technological capabilities? Their openness to technical solutions you might propose? How might your engineering expertise be an obstacle to enacting contextual listening?*

  c) *What kinds of questions have you asked them? What time has been devoted to listening and trust building? What role should listening play in future interactions?*

- *Map the location, knowledge, and desires of your design team or engineering homework group.*

- *Map the location, knowledge, and desires of those involved in the case studies in Chapters 6 and 7.*

5.9   CONCLUSION

Overall, we agree with the final bullet in the Barcelona Declaration, which calls for engineers to "Listen closely to the demands of citizens and other stakeholders and let them have a say in the development of new technologies and infrastructures" (Engineering Education, 2004). In this chapter, we described several dimensions of contextual listening to enable you to be better prepared to listen to and engage local community members' perspectives. Now that you are aware of the barriers to and benefits of contextual listening as well as PDS as an alternative problem-solving, listening-centered approach suited to SCD contexts, we hope you are in a more confident position to enact contextual listening. We also hope your SCD projects respect cultural pluralism and result in greater project success in the coming years.

Yet we also realize that many of the ideas in this chapter are abstract and can be difficult to apply, especially while one is hundreds if not thousands of miles from the actual partnering community. To help make many of these abstract ideas more concrete, the next two chapters present SCD case studies. In those chapters, you will encounter commendable practices of community-centered listening. We hope you are inspired by these exemplary engineers.

Exercise 44

- *Have each team member write the 5–10 most crucial questions to ask members of the local community. Then compare questions, using the Appendix as a catalyst to help select the 5–10 best ones overall.*

- *As a team, select all or some of the questions related to dimensions of contextual listening (see Appendix for a more complete list), if possible as they relate to a SCD case. Have individual team members write brief but anonymous responses to the questions associated with some or all dimensions. At the next team meeting, mix up the written responses, and each team member will take and read one that is not his or her own. Then switch again until you have read most or all of the other team member's responses.*

    a) What differences in perspectives emerges?

    b) What new insights, issues, or questions emerge?

    c) How might the team change its approach to future communication as a result of your discussion?

- *Have each team member draw a pie chart that estimates what percentage of the community your team has listened to so far via one-on-one conversations, small or large group meetings, or in other forums. Also have each member draw a second pie chart that represents your listening goal, in terms of what percentage of the community you will listen to prior to project implementation. Each member should note the steps necessary to move from the first to the second pie chart. Then as a team, compare pie charts and steps. As you discuss these, consider these issues:*

    a) Whose perspectives will ultimately be included and excluded? What benefits will arise from their inclusion? What risks are associated with their exclusion?

    b) The pie charts represent the quantity of community members the team has listened to, but what about the quality of listening? What kinds of listening help more than harm, and demonstrate the kind of follow through that shows you have listened and understood community perspectives? What forms of listening detract from or contribute to local community self-determination and ownership?

    c) Well-intentioned questions tend to only be effective in eliciting honest answers after building a foundation of mutual trust. How will this foundation be established?

    d) The questions we ask the community reveal much about our assumptions and values. What kinds of questions imply a deficit model? A capacity model?

    e) How might the PDS approach above be used in your listening process?

APPENDIX: QUESTIONS TO FOSTER UNDERSTANDING OF DIMENSIONS OF CONTEXTUAL LISTENING

Table 5.1:	
Dimension	**Questions**
Situated Context	• What are the origins of the community? What are the different ways to educate oneself to search and listen for these origins? • What keeps the community living together? What are their values and interests? • Why do/should they collaborate with each other? With your team? • How is the community organized? What are the differences (explicit and implicit) in power, status, privilege and wealth within the community? • What historical experiences has the community had with SCD projects? What are its attitudes toward outsiders? How might the cultural, political, and economic relations between your home country and the project site country shape local community members' (always diverse) attitudes about you and people from your country? How do the same relations shape your attitudes toward the local community? • How would you characterize the general community spirit--as hopeful, demoralized, and/or otherwise? What recent or less recent events have influenced this state of being? (Tools for exploring such questions appear in the description of PDS.) • As you listen, how do community members indicate whether issues of gender, culture, nationality, social class, and race/ethnicity inform the community's diverse outlooks on itself, SCD, and outsiders?
State over Trait	• What are the most salient traits of an effective listener? What are your strengths and limitations as a listener? • How has your engineering training affected your listening capabilities? • How can you constantly assess and reassess your own degree of openness to perspectives that differ from your own? How open are you to connecting meaningfully with people who hold differing worldviews and everyday perspectives? • How does empathy, the ability to see the world through someone else's eyes, fit within your worldview? How does empathy inform listening? • What opportunities and barriers exist to making effective, contextual listening a state of being you can enter (and exit) at will? What risks are associated with such a state?

Table 5.1:

Relational for Transformation

- How comfortable are you establishing positive personal and working relationships with people from other cultures? How much time have you spent living in another culture as an outsider and/or foreigner?
- What life experiences might facilitate or hinder your ability to meaningfully engage with and learn from people who may be less technologically "advanced" than you consider yourself to be? Or with people who may have a different religion, skin color, worldview, and set of life aspirations and expectations?
- What might help establish (or prevent you from establishing) relationships that result in a) identifying shared struggles and b) enacting community-supported transformation?
- How do you develop and maintain trust, the glue of effective relationships?
- What would need to occur for you to have the capability to not only listen to diverse community perspectives regarding an SCD project, but to also be able to re-articulate what you think you heard so the community members can check your understanding of their perspectives?
- After checking your understanding of what you have heard, how can you then apply it _in partnership with_ those who have spoken?

Recognizing Capacity

- What kinds of listening attitudes or activities might convey a deficit model mentality? Note in particular the risks associated with technological determinism described in Chapter 1.
- Since the deficit model is often imbued with a sense of cultural superiority, how can non-local SCD participants learn to listen to local constraints yet consciously focus on the community's capacities?
- If non-local SCD practitioners genuinely seek to enter the cultural frame of reference of local community members, what local knowledge, practices, and resources emerge as capacities?

	Table 5.1:
Self-Determination	• How can you "lay down" your potential contributions to an SCD project without forcing the local community to take them? A contextual listener sees a refusal to take offered contributions not as rejection but as an act fostering self-determination.
	• What kind of questions and listening can help you transcend the desire to help or empower others—which could replicate the deficit model and erode community self-determination—and instead, as Esteva suggests, to see if their struggles are also your struggles? And if they are, how can you work toward overcoming shared struggles collaboratively so as to foster self-determination?
	• What forms of listening detract from or contribute to self-determination?
	→ *Ownership:*
	• How did the project begin? Who initiated it? Who was included and excluded from that initiation process? Why?
	• Why was the project initiated? Who defined the project outcomes? Whose perspectives were (not) listened to in the outcome definition process?
	• How well does the project represent the cultural concerns, preferences, and practices of *all* local community stakeholders? And, if it cannot represent all stakeholders, how about those that were left out? Should they be compensated? Included in other ways?
	→ *Benefits:*
	• Whose concerns and interests does the project represent? Protect? Harm?
	• Who stands to gain from the project? How?
	• Who may suffer or be disadvantaged? How? To what extent do SCD projects give priority to students' learning and their international exposure (e.g., to make them more competitive in the job market) and the philanthropic image of sponsoring universities (e.g., to make them more attractive to prospective students and donors) over long-term risks and costs (time, money, resources, identity, etc.) to the communities?
	• Whose benefits remain central as the project unfolds?
Shared Accountability	• Who is accountable for the project's success?
	• How does this accountability shift over time?
	• What listening and other communicative actions and practices foster shared accountability?
	• How do your words such as "ours" and "yours," "we" and "you" create or bridge distances in this shared struggle?
Bias Awareness	• How can contextual listening serve as a way to counter known/suspected biases or unveil unconscious ones?
	• What are the best ways to avoid common biases (e.g., documentary, dominant and hidden voices biases, etc.)?
	• What role does and should the voluntary disclosure of recently recognized biases play in SCD community and project discussions?

Table 5.1:

<table>
<tr>
<td rowspan="3">Multiple Perspective Integration</td>
<td>• How does our view of the community's needs shift when we see how all community perspectives and sectors (economic, political, etc.) are layered and interwoven to form a complex, integrated whole?</td>
</tr>
<tr>
<td>• How do mechanistic compartmentalization of community perspectives and sectors contrast with holistic integration of those perspectives and sectors? How does each affect when, where, why, and how we listen?</td>
</tr>
<tr>
<td>• Why can the act of not accounting for the richness of human perspectives lead us to unsituated, decontextualized conclusions, actions, and decisions? How do we avoid this outcome?</td>
</tr>
</table>

RECOMMENDED READINGS

Achebe, C. (1972). Dead men's path. In *Girls at war and other stories.* London: Heinemann.

Adas, M. (2007). "Imposing modernity." In *Dominance by design: technological imperatives and America's civilizing mission.* Cambridge, MA: Belknap Press of Harvard University Press.

Botes, L. and van Rensburg, D. (2000) Community participation in development: nine plagues and twelve commandments. *Community Development Journal* **35**(1): 41–58.

Burkey, S. (1993). *People first: A guide to self-reliant participatory rural development.* London and New York, Zed Books.

Chambers, R. (1997). *Whose reality counts?: Putting the first last.* London, Intermediate Technology Publications.

Chambers, R. (2002). *Participatory workshops: A sourcebook of 21 sets of ideas and activities.* London, Earthscan Publications Ltd.

Chambers, R. (2008). *Revolutions in development inquiry.* London, Earthscan Publications Ltd.

Cornwall, A. and Jewkes, R. (1995). What is participatory research? *Social Science and Medicine* **41**(12): 1667–1676.

Kristof, N.D. and WuDunn, S. (2009). *Half the sky: Turning oppression into opportunity for women worldwide.* New York: Knopf.

Kumar, S. (2002). *Methods for community participation: A complete guide for practitioners.* Warwickshire, UK: Intermediate Technology Publications.

Mason, K. (2001). Participatory technology development. In *Brick by brick: Participatory technology development in brickmaking.* London, UK: YIDG Publishing. pp. 34–47.

Mortenson, G. and Relin, D. O. (2006). *Three cups of tea: One man's mission to fight terrorism and build nations…one school at a time*. New York: Viking.

Mortenson, G. (2009). *Stones into schools: Promoting peace with books, not bombs, in Afghanistan and Pakistan*. New York: Viking.

Participatory Development Forum (2005).
`http://pdforum.org/en/newsletters/`
A quarterly newsletter devoted to "deepen our understanding on participation, strengthen north-south and south-south dialogue, and provoke us to work in earnest to make participation more meaningful and more effective in promoting social justice."

Salmen, L. F. and E. Kane (2006). *Bridging diversity: Participatory learning for responsive development*. Washington, DC, The World Bank.

Schneider, J., J. Lucena, and J. A. Leydens. (2009). Engineering to help: The value of critique in engineering service. *IEEE Technology and Society Magazine* **28**(4): 42–48.

Slim, H. and P. Thomson (1995). *Listening for a change: Oral testimony and community development*. Philadelphia, PA., New Society Publishers.

REFERENCES

ABET (2004). Criteria for Accrediting Engineering Programs. Baltimore, MD, ABET, Inc. Engineering Accreditation Commission. 123

Adas, M. (2007). "Imposing modernity." In *Dominance by design: technological imperatives and America's civilizing mission*. Cambridge, MA: Belknap Press of Harvard University Press. 118, 121

American Society of Civil Engineers. (2004). Committee on Academic Prerequisites for Professional Practice, Body of Knowledge for the 21st century. Reston, VA. American Society of Civil Engineers. 123

Astin, A. W., H. S. Astin, et al. (2003). Leadership reconsidered: Engaging higher education in social change. Washington, DC, Project Kaleidoscope. 123

Bishop, R. (2005). Freeing ourselves from neocolonial domination in research: A Kaupapa Māori approach to creating knowledge. In *The Sage handbook of qualitative research*, 3rd Ed. N.K. Denzin and Y.S. Lincoln Eds. Thousand Oaks, CA: Sage. pp. 109–138. 131

Bridger, J. C. and Luloff, A. E. (1999). Toward an interactional approach to sustainable community development, *Journal of Rural Studies* **15**(4): 377–387, 126

Burkey, S. (1993). *People first: A guide to self-reliant participatory rural development*. London and New York, Zed Books. 121, 124, 139

Downey, G. (2008). Background on EPS and listening. May 14, 2008, email to J. C. Lucena. 141

Downey, G. and J. C. Lucena (2003). When students resist: Ethnography of a senior design experience in engineering education. *International Journal of Engineering Education* **19**(1): 168–176. 135

Downey, G. and J. C. Lucena (2006). Are globalization, diversity, and leadership variations of the same problem? Moving problem definition to the core (Keynote Address). Proceedings of the Annual Conference of the American Society for Engineering Education, Chicago, Illinois. 135

Downey, G. L. (1986). "Risk in Culture: The American Conflict over Nuclear Power." *Cultural Anthropology* 1: 388–412. 141

Downey, G. L. (1988). Reproducing cultural identity in negotiating nuclear power: The union of concerned scientists and emergency core-cooling. *Social Studies of Science,* 18: 231–64. 141

Downey, G. L. (1988). Structure and practice in the cultural identities of scientists: Negotiating nuclear wastes in New Mexico. *Anthropological Quarterly* 61: 26–38. 141

Downey, G. L., J. C. Lucena, et al. (2006). "The globally competent engineer: Working effectively with people who define problems differently." *Journal of Engineering Education* **95**(2): 1–16. 141

Easterly, W. (2006). *The white man's burden: Why the west's efforts to aid the rest have done so much ill and so little good.* New York, Penguin. 118, 121, 140

Engineering education in sustainable development (2004). Declaration of Barcelona. *International Conference*, Barcelona, Spain, 27–29 October. 142

GATE International (n.d.) Cross-section of a VSBK with single shaft, chain block unloading. Accessed January 31, 2010 from `http://www.basin.info/gate;vertical.htm`

Hagen, K. (2008). *Introduction to engineering analysis.* Englewood Cliffs, NJ, Prentice Hall. 135

Heshusius, L. (1996). Modes of consciousness and the self in learning disabilities research: Considering past and future. In *Cognitive approaches to learning disabilities* (3rd Edition). D.K. Reid, W.P Hresko, and H.L. Swanson (Eds.). Austin, TX: PRO-ED. pp. 651–671. 134

Jackson, J. T. (2005). *The globalizers: development workers in action.* Baltimore, MD, The Johns Hopkins University Press. 118, 119, 120, 121, 140

Leydens, J. A. and J. C. Lucena (2009). Listening as a missing dimension in engineering education: Implications for sustainable community development efforts. IEEE *Transactions on Professional Communication,* **52**(4): 359–376. 134, 135, 136, 137

Leydens, J. A. and Schneider, J. (2009). Innovations in composition programs that educate engineers: Drivers, opportunities, and challenges. *Journal of Engineering Education* **98**(3), 255–271. 123

Mason, K. (2001). Participatory technology development. In *Brick by brick: Participatory technology development in brickmaking*. London, UK: YIDG Publishing. 121, 140

Narayan, K. (1993). How native is a 'native' anthropologist? *American Anthropologist*, **95** pp. 671–686. 134

National Academy of Engineering (2004). The Engineer of 2020: Visions of Engineering in the New Century. Washington, D.C., The National Academy Press. 123

Ogundimu, F. (1994). Communicating knowledge of immunization for development: A case study from Nigeria. *Communication for development: A new pan-disciplinary perspective*. A. A. Moemeka. Albany, NY, State University of New York Press:219–243. 124

Ramaswami, A, J. B. Zimmerman, and J. R. Mihelcic. (2007). Integrating developed and developing world knowledge into global discussions and strategies for sustainability. *Environmental Science and Technology* **41**(10): 3415–3421. 133

Reid, F. J. M. and S. E. Reed (2005). The role of participatory listening in collaborative discourse: Insights from studies of engineering design teams. *International Journal of Listening* **19**:96–100. 136

Salmen, L. F. (1987). *Listen to the people: Participant observer evaluation of development projects*. New York, Oxford University Press. 121, 124, 139

Salmen, L. F. and E. Kane (2006). *Bridging diversity: Participatory learning for responsive development*. Washington, DC, The World Bank. 124, 139

Schneider, J. and J. C. Lucena (2008a). Sika Dhari's windmill: A sustainable community development case study, Colorado School of Mines, Division of Liberal Arts and International Studies. 141

Schneider, J. and J. C. Lucena (2008b). Building organizations and mapping communities: A civil engineer's work after discovering water, Colorado School of Mines, Division of Liberal Arts and International Studies. 141

Schneider, J., J. Lucena, and J. A. Leydens. (2009). Engineering to help: The value of critique in engineering service. *IEEE Technology and Society Magazine* **28**(4): 42–48.

Schultz, K. and L. Smulyan (2007). Listening as translation: Reflections on professional development work in a cross-cultural setting. *Learning Inquiry* **1**(2): 99–106. 129

Servaes, J. (1991). Toward a new perspective for communication and development. In *Communication in development*. F. L. Casmir (Ed.). Norwood, NJ, Ablex:51–85. 140

Shiva, V. (1993). *Monocultures of the mind: perspectives on biodiversity and biotechnology*. London, UK and Atlantic Highlands, N.J.: Zed Books. 118, 121

Shiva, V., Ed. (1994). *Close to home: Women reconnect ecology, health and development.* London: Earth-scan. 138

Slim, H. and P. Thomson (1995). *Listening for a change: Oral testimony and community development.* Philadelphia, PA., New Society Publishers. 121, 124, 129, 137, 138, 139, 140

Starosta, W. J. (1994). Communication and family planning campaign: An Indian experience. *Communicating for development: A new pan-disciplinary perspective.* A. A. Moemeka. Albany, NY, State University of New York Press: 244–260. 124

Wikimedia Commons (2009). Accessed Nov. 30, 2009 from `http://commons.wikimedia.org/wiki/File:El_Cajon_Dam_Honduras.jpg` 120

Wikoff, K., et al. (2004). Evaluating the communication component of an engineering curriculum: A case study. Proceedings of the ASEE Annual Conference and Exposition, Salt Lake City, UT. 124

CHAPTER 6

ESCD Case Study #1: Sika Dhari's Windmill

While writing this book, we discovered that there were not many ESCD case studies available that would assist students in thinking through the issues we've discussed in this book so far—the role that engineering students can play in SCD projects, the value of placing community at the center of any ESCD efforts, and the importance of listening. There are many cases dealing with large-scale development, or with the discoveries of anthropologists and sociologists; still others dealt with the impacts of development on communities. But the stories of the engineers involved in small-scale projects are rarely told, and almost never from a critical perspective that encourages self-reflection and analysis. Our hope is that the case study in this chapter, and in the chapter that comes after, will provide a start for thinking through some of the questions we have raised in this book so far.

We should also note that the case was constructed from an interview conducted with the professor who organized the ESCD project. Because of the way the actual case concluded, it was not possible to access or interview community members. This is a limitation of the case presented below—a stronger case would present the voices of engineers and community members together. Even with this drawback, however, we believe you can think through many of the issues presented in earlier chapters of this book in relation to the professor's experiences and reflections.

The case below and the case in Chapter 7 are based on actual events, though the names and locations have been changed to protect the anonymity of those involved.

6.1   CASE OVERVIEW

This case deals with the events surrounding a Sustainable Community Development (SCD) project implemented in Sika Dhari village in western India. Professor Rani Natarjan has teamed up with a non-governmental organization (NGO), the American Environmental Protection Agency (EPA), a group of her graduate students, and others to work with the villagers of Sika Dhari in designing and implementing a windmill. The windmill will be used to generate energy for powering flashlights in the village.

Professor Natarjan is committed to soliciting community involvement in the planning stages of the project. She and her students participate in a community meeting with the villagers, where the villagers demonstrate a significant familiarity with development projects, and communicate this knowledge and their desires to the engineering team, in some cases via interpreters.

In the implementation stages of the project, however, the team runs into some problems. The first deals with the windmill's commercial charge controller. The controller constantly threatens to fail, is expensive, and is difficult to manage and repair. The second deals with safety testing: the windmill has not been tested according to international protocols. Both problems are mitigated thanks to the involvement of one of the villager's residents, Anil Agarkar, who is a professionally trained electrical engineer from Mumbai, a large Indian city. He and his wife have settled in Sika Dhari as activists and organizers, and they have the technical knowledge needed to repair the windmill and ensure its safe operation.

Unfortunately, Anil's political beliefs create friction with Professor Natarjan and her team: raised in a radical Gandhian tradition, Anil does not believe in intellectual property rights and is fairly suspicious of NGO's, governmental organizations, and of foreigners. As a result, he refuses to give credit to Natarjan's design team in publications about the project. Anil's activism has also aroused the attention of governmental officials in India, and so he cannot act as a legal spokesman for the village. This means that the engineering team struggles to transfer funds to complete the project and has difficulty getting a contract signed that would relieve them of liability in the project. In the end, Professor Natarjan leaves the project and cuts off communication with Anil and the village. She believes that the windmill continues to operate today.

How can we make sense of this story? How did the windmill come to be in the first place? How could two engineers come to hold such different political perspectives and engage the same community in different ways?

6.2   CASE SPECIFICS

6.2.1   INTRODUCTION

In 2004, Professor Rani Natarjan and a team of engineering graduate students from a US university traveled to a rural part of Western India to begin work on a sustainable development project. Natarjan had been in contact with a non-governmental organization (NGO) called India Now, whose mission is to organize volunteers interested in just and sustainable development projects in India. India Now had told Natarjan about a group of small villages in India interested in having a water and sanitation project installed. Natarjan believed that she and her students could help.

Key Term

NGO: Stands for Non-Governmental Organization. These are organizations that typically (though not always) operate outside of governmental organizations and are frequently non-profit. NGOs gained substantial influence in the 1980s, and they are frequently thought of as "filling in the gaps," providing services and goods where government programs don't or can't reach, though more recently they work hand-in-hand with government programs, making it sometimes hard to draw a clear boundary between the two. More and more engineers doing SCD work find it necessary to work with NGOs to ensure local support for their projects.

6.2.2   BACKGROUND

Natarjan, a civil engineer, was originally from India, and attended one of the Indian Institutes of Technology (IIT). IITs are the most prestigious engineering schools in India, recruiting the top 2% of all college-bound students in that country. After studying chemical engineering at the IIT, Natarjan pursued masters and doctoral degrees in civil engineering at a top research university in the US, where she developed in-depth expertise in water treatment modeling and life cycle analysis. As a graduate student, Natarjan was able to question the validity of the mathematical (stochastic) models that her advisor was applying to water resources and ended up shifting her attention to experimental research. At the same time, she ventured into a graduate course in cultural anthropology, which challenged her to apply ethnographic field methods and interviewing techniques in a soup kitchen that provided food to homeless persons.

After graduation, Natarjan began teaching at an American university. As her career progressed, she became interested in sustainability, and hoped to use her engineering expertise to help alleviate some of the conditions of extreme poverty in India. She was also interested in the learning opportunities sustainable development projects might afford her graduate students. The graduate students themselves were in school to work with Natarjan on research projects related to sustainability, and they were motivated to work with the villages in India. Many took it on themselves to learn Hindi, one of the major languages of India, before they departed.

Professor Natarjan and her students were not sure what to expect once they arrived at the villages. As a civil engineer, she assumed that their work was going to be related to water treatment. One thing Natarjan *was* sure of was that she wanted to educate herself and her students about how best to involve the Indian villagers in designing and implementing the disposal systems. In preparation for the trip, Natarjan consulted numerous documents on effective development practices, including documents from the World Bank and from an energy institute, a New Delhi think-tank devoted to sustainable development practices. Her previous training in anthropology and experiences in listening to the perspectives of homeless persons had sensitized her to the need for tools that would bring people's voices to the table. Both sources suggested a number of "best practices" developed by those working in the field for many years. Professor Natarjan skillfully combined the following practices with her previous knowledge and skills in ethnography:

- **Hold village meetings**. Natarjan firmly believed that meeting with the villagers to determine *their* desires for the project was important for its long term success.

- **Ask open-ended questions**. This would allow villagers to openly share their knowledge and desires. In particular, Natarjan wanted to begin by asking the villagers to tell her and her students the "good things about the village" first, to build trust, get a sense of the social dynamics there, and make a point to her students and villagers that the community had much to offer. Then, she asked if there was anything the villagers would change about their lives, and what she and her group could do to help. She tried to avoid asking "what do you need?" for she understood that this question implies a "lack" or "deficiency," which could be interpreted as insulting the villagers and convey to her students the misleading idea that poor communities have few capacities.

- **Ask villagers to draw a "community map."** This strategy invites villagers to draw a map of their village that indicates the villagers' relationship to their environment and to each other. Maps can also indicate who is considered a part of the village (and who isn't). Such a strategy can reveal interesting dynamics not immediately obvious to the observer and can also be one way to overcome a language barrier (for a more intricate method of community mapping, see Chapter 7).

- **Observe**. It may seem obvious, but taking the time to carefully observe one's surroundings and the dynamics of a particular community is important. Professor Natarjan may not have always understood what she was seeing, or she may have filtered it through her own experiences, but at least she had more information that she could use later for reflecting on design processes, and she could use her observations to formulate questions.

- **Shadow**. Requesting permission to follow a villager around for a period of time (several hours to several days) can reveal information about how daily life actually unfolds in a community, again revealing unspoken relationships and procedures to the observer. Observation and shadowing are important checks to oral communication where villagers might tell outsiders what they want to hear.

- **Speak to the women separately**. In many communities (if not all), gender shapes communication practices and women's voices are often silenced. Speaking to men and women in separate groups, and then together, may reveal different needs and desires not expressed in mixed settings.

- **Listen**. All the practices above require careful and respectful listening to others' verbal and non-verbal languages. Natarjan had developed this ability during her ethnographic experience as a graduate student (for a detailed account on how to listen to community, see Chapter 5).

Exercise 45 *Develop a list of principles for community-centered ESCD work, based on your reading of previous chapters. How do Natajan's practices align with your list of principles? Do her practices add anything new to your list of principles?*

Equipped with this sensitivity and toolkit of strategies, Natarjan and her students departed for India. They hoped to perform an assessment using the strategies outlined above and to work with the villagers to develop a project that would be useful and sustainable. The next parts of the project are presented in terms of two phases—Participatory Planning and Implementation—the phases the engineering group went through as they worked with the village in developing and completing the project.

6.2.3   PHASE ONE: PARTICIPATORY PLANNING

Professor Natarjan and her students arrived in India and traveled to one of the villages, Sika Dhari, to begin their site assessment. India Now had arranged for contacts from the village to meet the group and introduce them to the village and its inhabitants. These contacts were Anil Agarkar and his partner Amani. Natarjan and her students were soon to learn more about this unusual couple and their relationship to the village. These "gatekeepers" would be central figures in the development of the project.

Natarjan and her students knew that it was important to understand the terrain and history of Sika Dhari. Being from India originally, Professor Natarjan knew that the village was populated by "tribals," and that this had significance for how the village was organized and developed. Indian society has historically been organized by caste, with people falling into a rigid hierarchy of classes indicating religion, wealth, race, and so on. Tribals, on the other hand, are typically not considered a part of the caste system. Rather, they often function in groups that are agriculturally and economically self-sustaining, and they are separated from the rest of Indian society by language, geography, religion, and economic organization. Their culture, environment, and economic systems have historically suffered at the hands of colonialism and industrialization, and as a result, tribals are sometimes afforded special recognition and protection by governmental bodies. See Figure 6.1 for a map of where tribals are located in India. To those coming from rich countries, tribal villages seem to be extremely poor and isolated, though that is changing due to increasing migration and improved transportation.

Sika Dhari is home to approximately one hundred tribals, though the number fluctuates as young people leave to work outside the village or to get technical training or other education. The village is located in a hilly part of Western India, known for heavy rainfalls. Because of the terrain, rainwater often washes downhill, eroding soil and making agriculture difficult. With the help of Anil Agarkar and Amani, the village has been working to develop "bunds," earthen berms and dams that will decrease erosion and improve harvesting, pictured below in Figure 6.2.

The villagers themselves, therefore, were not new to aid projects. India Now has funded several projects in Sika Dhari and the surrounding villages, and Anil and Amani have worked to organize women's collectives, which have led to the creation of micro-lending programs, food-for-work programs, and increased cash flow in the village. Many villagers have at least an elementary education, and several have received technical training elsewhere. The village is also familiar, however, with less successful development projects that have taken place in neighboring villages; news travels fast in this area, and villagers are not naïve when it comes to interacting with outsiders.

Again, Natarjan initially assumed that her team would be involved in a water-related project. Anil and Amani arranged for Professor Natarjan and her students to participate in a community meeting about the village's plans for building toilets. All major decisions taken in the village must be agreed upon by consensus; the bunds projects, for example, often take six months to a year to plan because consensus is created slowly and carefully, and must be complete. Decision making by consensus often frustrates those that value efficiency and economy in the use of time, but Natarjan

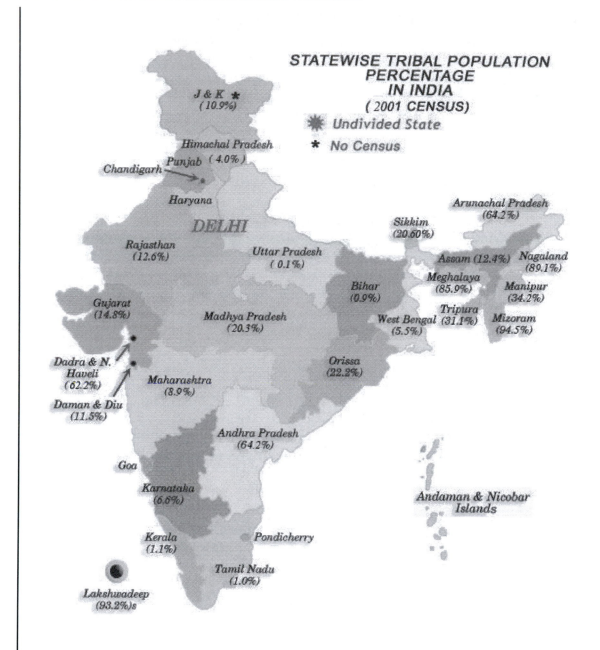

Figure 6.1: Tribal regions of India (in purple). (Source: http://www.originalworld.com/images/ public/maps/map_india_tribal.jpg).

Figure 6.2: Making bunds for a rice field, Warangal District, Andhra Pradesh, India. (Source: http://farm1.static.flickr.com/11/16434553_b7d12a15ca.jpg?v=0 Credit: Glenn Davis Stone).

knew that the engineers would need to allow this decision-making process to unfold at its own pace if the project was to have long-term success. The engineers learned that they would be participating in one of these meetings, and they were ready to use the list of open-ended questions they had prepared in advance. Natarjan would also be shadowing one of the woman villagers during her stay, to learn more about her life.

Professor Natarjan spoke Hindi fluently, and some of her students were beginning to speak it as well. As a result, they could communicate with some members of the village, though many villagers spoke a tribal language, rather than Hindi. Because of the language barrier, the engineering group relied heavily on Anil and a few of the village youth to translate into Hindi or English. Communication was sometimes slow and difficult, mediated by Anil and others. At the community meeting, the villagers told the engineers how proud they were of their village: they had thriving farms, food stocked for a year, and homes that they had built and maintained. They also had a growing lentil processing business that permitted them to earn some cash and purchase cows, providing milk and other income.

When the conversation turned toward what the villagers would like to see changed, the villagers spoke about their desire for toilets. They had been walking into the forest at night to relieve themselves, which was inconvenient and potentially dangerous. The engineers learned that the community members knew an extensive amount about the options available to them when it came to installing toilets—they had seem similar projects implemented in surrounding villages, or they had learned about various technologies in their technical training. Biogas toilets, in particular, are a common site in India and Nepal (see Figure 6.3). The conversation proceeded something like this:

Professor Natarjan: Well, how about a community toilet?

Villagers [laughing openly]: No! Definitely not. Who will clean the toilets? Who will maintain them? No.

Professor Natarjan: What if a group of houses shared a common toilet among them?

Villagers [laughing, shaking heads]: No, that won't work either.

Professor Natarjan [smiling]: Why not?

Villagers [still laughing]: Because then father-in-law will know when daughter-in-law goes to the bathroom! How odd!

Professor Natarjan: Well, then. How about bio-gas toilets?

Villagers [laughing again]: They don't work.

Professor Natarjan: How do you know they don't work?

Villagers: We've seen those biogas units in other villages; we know where all the leaks are.

Professor Natarjan: Well, what then?

Villagers: How about ceramic toilets? Like in the city? We like those.

The conversation about toilets went on for some time. Laughter became a sort of cue for Natarjan, signaling that the villagers not only knew what they wanted but would politely let her know that her proposals were comical. In the end, two families in the village did decide that, down the road, they would try bio-gas toilets, despite all of their reservations. For now, the toilet project would be tabled because, the engineers would learn, the villagers had a project that they found more pressing.

As the community conversation developed, the engineers learned that the villagers would like a renewable power system that could charge flashlights. Many were uncomfortable having to walk through the forest in the dark, and they felt that some form of electricity would benefit them. However, the village was wary of solar power because they had heard of failed solar projects in abutting communities. They were also curious as to how the electricity would be allocated and how individuals would pay for it, a question the engineering group had not even considered.

The community and the engineers went through another long conversation in which the villagers educated the engineers about which technologies would and would not work. So far, no traditional engineering problem solving had been deployed; no conceptual design had been made. Villagers and engineers were deeply involved in defining a problem in the midst of complex politics.

Figure 6.3: Public biogas toilet in Tamil Nadu, India. (Source: `http://southasia.oneworld.net/ImageCatalog/toilet-in-musiri.jpg`).

Natarjan and her team discovered that the villagers strongly favored a wind power solution as a response to a hydroelectric dam that was being constructed nearby and was going to flood lands occupied by neighbor rural communities. The villagers of Sika Dari and an activist group associated with the village wanted to make a political statement against the dam by choosing an alternative source for their energy needs. Protests against dams, as pictured in Figure 6.4, had been occurring in many parts of India.

Clearly, technical choices are shaped by politics that sometimes are difficult to understand. As Natarjan puts it, "We learned that there's all these power things [politics] going on…we found out that other communities didn't even prioritize electricity [water was more critical], you know, and it turns out that it may be a political thing with the group that is against the dams wanting to show that there's potential for other forms of renewable energy…."

In the end, the community reached consensus that they would like to try a small windmill. The windmill would charge large batteries that could be used, in turn, to power small batteries for flashlights. The engineers agreed that this would be a project they could pursue in spite of their lack of experience in electrical engineering projects.

It is important to know that wind power has become increasingly more common in India over the last twenty years. Nearly 2% of India's electricity comes from wind power, which seems small, but is comparable to the relative amount produced by other countries invested in this technology. In fact, India has the potential to become a major world player in wind power, competing with Germany, Spain, the US, and Denmark, and windmills can be found in many of India's states. India

Figure 6.4: A protest against the flooding caused by the building of dams in rural India. (Source: `http://www.aidboston.org/MedhaPatkar2009/images/MedhaPatkar_protest.jpg` Permission pending.)

is posed to develop significant capacity in wind generation in the coming decades, though it will probably continue to provide a small amount of overall electricity generation in the coming years. That said, many windmills in India are quite small compared with commercial windmills. As you will see below, one of these small windmills was eventually built for Sika Dhari. More information about the type of windmill developed in this case can be found at `http://greenwindenergy.net/ruralindia.html`.

As the community meeting disbanded, Professor Natarjan commented that she was struck by the level of consensus-building, wisdom, and mature engagement that she had just witnessed. She commented: "They were self-sufficient. They had food for one year; they showed us, you know, how they stored their food. And you know, they seemed wiser, and they knew, you know, like I said, their knowledge base was quite high, and they were I thought, we all thought, that they were much more democratic than any of us were."

Exercise 46 *Think about the principles for contextual listening developed in Chapter 5. In what ways does the preceding section on "participatory planning" exhibit principles of contextual listening? What information is still missing? What challenges have they encountered so far? What more does the engineering group need to know or understand before returning to the United States to work on the project?*

6.2.4   PHASE TWO: PROJECT IMPLEMENTATION

The engineering team spent several days in the village, getting to know Anil and Amani, shadowing some of the villagers, observing, and thinking about the windmill project. All were eager to return to the United States to begin work on the design.

Upon their return to the US, the group eventually settled on a wind turbine design by an expert on small wind energy designs. They also enlisted the help of an electrical engineer from Professor Natarjan's university, and a professor from an Indian university near Sika Dhari. Anil and Amani and the villagers were also consulted and trained via a series of workshops run by Natarjan's graduate students and the Indian professor. At last, the turbine was erected in the village, and Natarjan and her students were pleased to learn that the villagers would have energy to charge flashlights to guide them through the night in Sika Dhari.

Unfortunately, there were some problems with the design. The main glitch was with the *charge controller*. The original design called for a commercial charge controller, which was expensive and complex, and which began to fail almost immediately. If it failed in Sika Dhari, it would be nearly impossible for villagers to replace or repair it since this artifact was imported. All support and maintenance for the controller resided elsewhere. The engineering group was eventually able to design a more appropriate charge controller, one that was simpler and easier to control, but it has not yet been implemented in the village. Natarjan and her team also had their concerns about who might manage the controller. Fortunately for Sika Dhari, Anil—the team's main contact in the village and an electrical engineer by training—was able to manage the charge controller for the time being. But the engineers were well aware that not every tribal village had an "Anil" to help them. Finding highly educated urbanites in a remote tribal village was surely unique. As Natarjan put it, "It's very rare, very unusual. He [Anil] is probably literally one in a million."

Of greater concern to Professor Natarjan and her team was *safety testing*, an issue that worries Professor Natarjan to this day. Windmills are supposed to be safety tested according to International Electrotechnical Commission (IEC) standards, which include power performance, noise measurement, and blade structural tests, among others, but small, hand-made windmills have never been tested, although they are installed routinely all over the world. After the windmill was erected, Natarjan contacted the Environmental Protection Agency (EPA), which had supported the project along with India Now, and informed them that she was concerned that Sika Dhari's windmill hadn't been tested. Natarjan and the EPA agreed that because the windmill was being monitored by Anil, a trained engineer, it would be allowed to remain without testing (for IEC safety standards, see `http://www.awea.org/standards/iec_stds.html`).

But Natarjan still had her concerns. "If the battery overcharged for a second," she said, "it would make the wind turbine go up and short circuit. And if the winds are very high, you know, [the windmill] can spin like a propeller; it's very dangerous, and people say the blades can fly several kilometers [if dislodged]. It could be dangerous." But testing at this point in the project was no longer feasible. Testing at a laboratory in the US could cost up to $100,000, while the cost of the entire project itself had only been $1500. Furthermore, many of the IEC's protocols were established to test large megawatt generators, not small windmills like the one in Sika Dhari. For example, protocols often call for hundreds of hours of testing at various wind speeds and altitudes; such testing was beyond the scope of the project.

Natarjan and her group were also concerned about *liability*. If something were to happen to the windmill or to the villagers as a result of the project, Natarjan and her group did not want to be held accountable. They had created a contract before the workshop, which they wanted Anil to sign as village representative, which would indicate the scope of their responsibility to the project. However, none of the Indian parties signed the contract. After the windmill was installed, the relationship between Natarjan and Anil became strained.

Anil Agarkar and his wife Amani had originally moved to Sika Dhari in the early 1990s. They were both political and social activists, committed to a Gandhian philosophy that emphasized the decentralization of wealth, anti-materialism, the preservation of tribal cultures and ways of life, the liberation of lower-caste groups, and the democratization of government and resources. Anil and Amani were particularly concerned by the environmental degradation of tribal areas, such as the one where Sika Dhari is located. This area is the site of contentious struggles over dam-building, watershed protection, and industrial pollution of waterways. The couple had been particularly active in protesting such degradation, and they had at times run afoul of authorities who resented their opposition.

Anil and Amani's activism made *communicating* with the village about the wind project complex. In retrospect, Natarjan suggested that it was often difficult to know exactly whose interests were being represented in community discussions: "They [the activists] want to show that hydro is not the only thing but they're not the villagers…[so] 'Who's representing whom?' It's very hard to make out up front you know, you may work with the best organizations but what's their internal politics?" Even many months after the project was completed, Natarjan was still not sure she understood the decision-making process in the village, nor the role of various NGOs in the windmill project.

Professor Natarjan found that Anil was unaccustomed to acknowledging the work of the windmill designer or the rest of the design team in newsletters he sent out about the project; nor would he acknowledge EPA funding, perhaps because he was wary of US governmental organizations. In Gandhian fashion, he also resisted the idea of intellectual property, feeling that all ideas should belong to the people. "What we find," said Natarjan, "is their politics is almost so difficult that you couldn't take any steps forward because they don't want to acknowledge foreign support even if they get it." Anil's status as a political outsider in India also made funding the project difficult because he could not officially register to receive money on behalf of the villagers for the project. And finally, he was reluctant to sign the group's contract. He did sign an informed consent form stating he understood the risks of the technology, but the form was accidentally left behind by a graduate student who was visiting the village for a second time.

Exercise 47 *Compare this SCD project with the one at the beginning of Chapter 3. What could have the students in that humanitarian engineering project done differently to make their project be like Natarjan's? Or perhaps could Natarjan have something to learn from the project in Chapter 3?*

6.3 CONCLUSION

Given the difficulties with the contract and with Anil, Professor Natarjan is no longer in contact with the villagers of Sika Dhari. She believes that the windmill is indeed up and running, and that the villagers see the project as a success. But we do not know much about how the villagers use the wind power. Natarjan regrets this sequence of events because she knows that other villages around Sika Dhari were interested in similar power or water supply projects, and she had hoped to return to work with them. Those plans have been abandoned.

For her part, Natarjan has gone on to plan a new wind project in a neighboring country. Before she even departs for her site visit, however, the project's windmill will sit atop a mountain in Colorado for hundreds of hours, undergoing testing for safety. She has met some resistance on this point from those who say, "Oh, but the one in India is working!" She responds, "Yeah, but it's only working because the guy there is exceptional. And that is absolutely not common. And he [Anil] even said that but for [his] quick action…there were times when it was [dangerous]." Natarjan is not willing to undertake this risk again.

Natarjan also insists that she will have her contact in a neighboring country sign a contract before installation even begins. Still, Natarjan harbors no hard feelings toward Anil and Amani; she finds their work and passion extraordinary. But she does wonder about exactly whose voices she heard in the community meetings in Sika Dhari. "It's hard. The politics of the thing were very, very difficult. We wonder, 'Who is the speaker? Who is the voice?' And it's never been clear to me because we can't understand the tribal language."

Though excited about her new project in the new country, Professor Natarjan has also begun to devote significant energy to urban sustainability projects in the United States, in her home city. There is a part of her, she indicated, that questions the feasibility of sustainable development projects abroad. She wonders, provocatively, who in fact benefits more from these projects—the villagers or the students who are sent there? She says now, "What I found is people in the villages are smart, they know what's happening, they know what they need. They may not have funds to do certain things that they want to do, but you know this whole thing of going and doing all this is actually benefiting our students more [than the villagers] because it's opening their eyes. So let's be honest and say 'Yeah it's a good international exposure for our students but do we contribute that much to these communities?' I don't know. I don't know. I seriously don't know….I still wonder if [we] left [the villagers] alone, if they would be just fine."

6.4 QUESTIONS FOR REFLECTION

1. After reading this case, what are your feelings about ESCD? Are such projects worth the time and effort? What lessons do you, personally, draw from this case?

2. What positive outcomes emerged as a result of this particular ESCD project? Were there negative outcomes?

3. What criteria do you use to assess such a project?

4. Describe Professor Natarjan's strategy for designing and implementing the project. Why does she do things the way she does? How would you characterize her approach? Is it successful?

5. How would you describe the villagers of Sika Dhari? How would you describe the members of the engineering team? How would you describe the interactions between the two?

6. What lessons came out of the engineering team's interaction with the villagers at the community meeting?

7. Where do the technical dimensions of this project begin? Where do they end?

8. Who are the experts in this case study? What makes them the experts?

9. Consider the following dimensions of sustainability. How "sustainable" was the project, according to these different dimensions?

   - **economic sustainability**: the project should not tax the village's monetary resources, should be implemented within the constraints of the project budget, and that should be maintained over time

   - **environmental sustainability**: the project should not harm the environment or tax natural resources unnecessarily

   - **social/cultural sustainability**: the project should grow out of the villager's desires, and be "owned" by them, literally and figuratively

   - **technical sustainability**: the design of the project should be feasible, safe, and sound, and should function effectively for its expected lifespan. The technology should be appropriate to the community's location, knowledge and desires

10. In the conclusion of the case, Dr. Natarjan suggests that she has mixed feelings about ESCD projects abroad. Discuss and analyze this conclusion in light of the facts of the case.

CHAPTER 7

ESCD Case Study #2: Building Organizations and Mapping Communities in Honduras

7.1   INTRODUCTION

We hope by now the central importance of incorporating communities and developing contextual listening in engineering work for SCD is clear. But you may be wondering how this idea looks in real practice, by real engineers, in the real world. This case study is an abbreviated history of a civil engineer who effectively incorporated communities as a central part of her work. To that end, her pathway was not easy or straightforward, as yours probably will not be. Through many events and circumstances, some of which were outside of her control, she gained the knowledge and experience to understand, value, and work with community.

Most importantly, as a lesson for those committed to community development, she conceived and implemented strategies that empower communities to take control of their own water consumption, sanitation, and treatment. Through this case study, we hope to show that despite the huge challenges posed to engineering in earlier chapters, engineers—in collaboration with many other groups, such as community members, NGOs, government officials, etc.—can successfully engage and empower communities to take control of their own destiny.

7.2   BECOMING AN ENGINEER

Elena Rojas[1] was born in Honduras' capital city, where her father was studying to become a teacher. She spent most of her childhood, though, living in a small town in Honduras, where her family had moved so that her father could work for a large, multinational fruit company. In high school, where students in Honduras have to specialize early for professional work, Rojas wasn't sure what she wanted to study. At first, she chose accounting because she liked math and wanted to be prepared for work in a profession. Accounting didn't satisfy her, though. Rojas felt a need to train in a profession where she could help people and in some way give back to her fellow Hondurans.

> You know I […] do not come from a very rich family. My parents, with a lot of effort, they sent us to school and then to high school and then I had the opportunity to take

[1]The names of persons and organizations have been changed to protect participants' confidentiality, promised to them as part of our ethnographic research on ESCD. Yet the stories, quotes, dates, and other information presented here remain factual.

my studies at the university because I was working all the time. So […] I really have to consider myself very lucky because very few people have the opportunity I have had. […] I know there are a lot of needs. And I need to work, that's all. I need to give something that I have got [….] I need to leave something. And that is simply what I did.

As a result, when Rojas began her studies at a large Honduran university, she first studied law. After a long process of analyzing the different career paths offered by the university, however, Rojas finally settled on civil engineering. She graduated in 1986, the first member of her family to receive a degree in engineering.

Even though Rojas had focused her studies on civil engineering, the education offered at the university then was very general: students took basic courses in sciences and mathematics, and for the most part, there was little specialization. Rojas took courses in mathematics, physics, structures, hydraulics, hydrogeology, water supply, sewage, and sewage treatment. She also studied mechanics of soil, roads, pavement, and other structures—mostly, a typical curriculum for a future engineer. But Rojas does remember one course that was different. Her water supply systems course was taught by a dynamic professor who used methods that were unusual for that time. He actually took his students out of the classroom and into the field, where students identified local water sources, walked the lengths of pipe, identified tank locations, and walked through local communities to understand how houses were distributed. Rojas clearly remembers the field exercise in this course:

> [It] consisted in designing a rural water system but didn't have contact with the community at all. It was just the technical part of going to the field, identifying the water source, walking all through the pipe, the location for the pipe line, the tank location, and we went through the community just to have an idea how the houses were distributed…we didn't talk to anybody. So our professor gave us the topographical survey and some guidelines to design the rural water system. That was all.

While it was unusual and forward-thinking for her professor to at least take them into the field, Rojas noted that she and her engineering classmates were learning how to design a technology to move and store water from point 1 to point 2, efficiently, without communicating with members of the community (see Figure 7.1). She would remember this gap between technical design and the importance of talking to people years later when she began her work as a coordinator for an non-governmental organization (NGO), an organization that mapped communities by establishing social relations first in an effort to provide water and sanitation systems.

Exercise 48 *Think about how Elena's experiences might be similar or different than yours in one of your engineering courses requiring fieldwork. Was your professor's approach purely technical? Or did it have social dimensions such as engaging communities or groups of people who might be affected by the science or technology being studied or who might have had important local knowledge?*

Rojas worked her way through the university system, supporting herself for her first two years of study as a high school math and physics teacher, and for the last two years as a consultant and

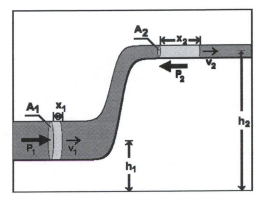

Figure 7.1: Engineering conceptualization of fluid flow through a pipe. (Source: http://www.4physics.com/phy_demo/fluid-flow.gif).

technical designer at a company that designed metal infrastructures. Rojas would go on to work for that same company as a technical supervisor for the three years following her college graduation. She gained more experience working in the field in her position there, but that experience was primarily limited to working with other engineers and managers and communicating technical information to her industry clients. She never communicated with the workers of the factories that she was helping build—there were no structures in place or impetus for doing so.

However, Rojas's life was about to take an interesting turn, a turn that would give her an opportunity to have great impact on how the Honduran government, NGOs, and citizens would understand and manage water as a resource in that country. While she was working for the metal infrastructures design company, Rojas ran into her old hydrology professor, who asked her if she was ready for a career change. Intrigued, Rojas asked for more information. The professor wanted Rojas to come and work for the municipal agency in charge of water (Division Municipal de Aguas, or DIMA), as a technician in the design department. DIMA was the municipal government agency with the mandate to administer water supply in the municipality of San Pedro Sula, the largest industrial and commercial area in Honduras (pop. 527,000). Rojas jumped at the chance because she knew working for the public agency would open doors for her:

> I think I could have had a better chance to get a better salary in the private sector but I wanted [to work in] water and sanitation […] in those topics. So I really liked that area. And secondly I wanted to get a master's degree and I knew that I could have a better chance by getting a grant if I was working in the public sector than in the private sector…I have to get a higher degree so I can really compete [with men] in the labor market […].

Rojas would end up working for DIMA for the next nine years, and would indeed complete her master's degree, which gave her legitimacy to work and speak in a male-dominated field.

7.3   DISCOVERING WATER

During her first years working for DIMA, Rojas was accepted as a student at a Northern European university, which she attended for two years. While there, she studied environmental sanitation. Although Rojas was also accepted to study at a university in South America, she chose to go to Europe so that she could learn English: "Even though I knew that I was going…to learn a lot of very high technology stuff that probably I would never apply to my country, […] I knew I had the opportunity to learn English because I was going to be somehow forced; the whole master degree was in English. So I made my decision thinking of that […] I needed to learn English." Her commitment to and knowledge of community would develop later. For now, she was gaining new knowledge about water as a resource that could be polluted and hence protected.

Rojas's decision to study environmental sanitation in Europe would end up being a fortuitous one, because it trained her in environmental sanitation, a field that was virtually unknown in Honduras. This training would have tremendous impact on Rojas's future career path and how Honduras would end up managing its water resources:

> You cannot construct a water system even in a rural area if you are not thinking of the [watershed] protection. You cannot build sewage systems if you are not thinking of how to treat those sewage […] for reducing the environmental impact on the rivers. Everything, especially in my country, everything you do impacts negatively or positively to our water resources. So I am very glad that at the end I could match my civil engineering with environmental sciences.

This training enabled Rojas to see water as a resource to be protected through policy instead of as a physical object to be moved from point 1 to point 2 through engineering. When she returned to DIMA, she brought this knowledge to bear on sanitation policy in Honduras. "I was very motivated because when I left DIMA in 1989, we didn't have any policy about sewage management. So when I returned to my country, I came back to work for DIMA and I started a very difficult process because I was the only one aware of the damage we were doing with the sewage in the city …." In the late 1980s, Rojas was particularly concerned with the large number of *maquilas* and other industrial factories in this particular area of Honduras.

Rojas felt that she had to use her education to affect what was happening with pollution in that area: "So immediately when I finished my first year in [European country], I said, 'I need to go back to my country and start […] to convince decision makers that we need environmental regulations,' because we didn't have anything at that time." As a result of her training abroad, Rojas was seeing water in a different way (see Figure 7.2).

Unfortunately, Rojas met with apathy, if not resistance, when she returned to DIMA to lobby her cause with the organization's management. Her peers could not see what she was then seeing. The manager at DIMA "never understood" what she was trying to tell him, and individual municipalities did not respond. Nothing moved forward until Rojas learned that the city where she worked was developing a sewage master plan funded by a loan from the Interamerican Development Bank (IDB)

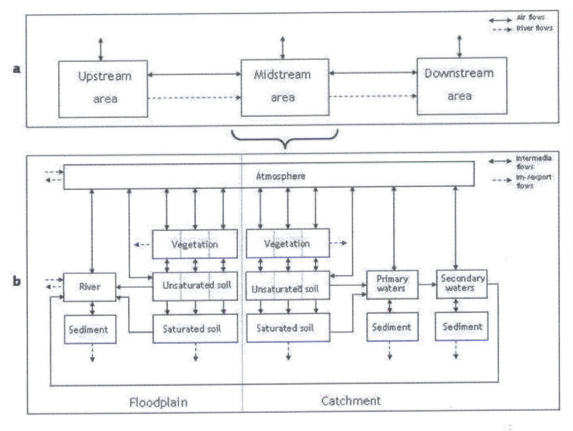

Figure 7.2: Conceptualizations of water in environmental science are complex and often include relationships among water, atmosphere, vegetation, and soil at different stages of a river flow. (Source: Radboud University Nijmegen).

and private funding from an Italian civil engineering consortium. She approached the coordinator of the plan and asked him about effluents:

> So I went to the person who was coordinating that program, and I told him, 'Okay, you are going to develop a master plan on sewage. What are you going to do with regulation?' And he said, 'No. We are not thinking about it.' [And I said,] 'So we need to know what kind of effluents this city is producing. What are you going to do about...' [He said,] 'No. We don't know yet.' So I offered [...] to make a characterization of all the effluent disposal places. And after maybe three or four months, they gave me a very little funding to do that. And I did it .... So it was very interesting because after struggling with many people, we finally created the regulation and we created within DIMA, (even though DIMA was only [...] created for the operation of the water and sanitation services of

the city), we created a water resources department as a result of the regulation, the rough regulation we created. And that happened in the year 1994....

It is important to understand that Rojas' new way of seeing water allowed her to introduce her colleagues to new concepts such as water quality, pollution, and sanitation. Rojas was carefully and successfully shaping DIMA's actions on water supply and sanitation, and she was garnering support and organizing people to make these changes happen. In one sense, she was a sort of water "revolutionary" in Honduras.

Exercise 49 *Think about how Elena's working experiences might be similar or different than yours during your internships, co-ops and/or jobs. Who has more authority to speak and perhaps change the organizations where you have worked? Men? Women? Those with advanced degrees? Those with interdisciplinary knowledge that includes science, engineering, humanities and social sciences? What do her and your experiences tell you about the relationship between having certain kinds of knowledge and having the ability to change organizations?*

7.4   CHANGING AN ORGANIZATION

Ten years earlier, in 1983, DIMA officials had created a "watershed protection unit," but it was unclear what this unit was charged with—its actions were largely unorganized and asystematic. Rojas brought back from Europe the idea that water is a "resource," and as a result needed to be managed and protected in new ways. What Rojas and a few sympathetic colleagues at DIMA did was to take this watershed protection unit and bring it together with hydrology, water control, and effluent control units, under the new name of Departamento de Recursos Hidricos (Department of Hydrologic Resources). This organizational reconfiguration was quite significant because "unit" was elevated to "Department" and "water" elevated to "resource." She brought people of different backgrounds and concerns together, encouraged them to talk to one another, and began to get things done. Making water a resource presents new challenges. On one hand, it is a welcome change for it elevates water as an area of official concern that deserves to be protected. On the other hand, it could make water the target of privatization, which would create inequalities among those who can afford to buy it and those who cannot.

Rojas notes that as an engineer she was not "supposed to" be building or bridging units the way she did in DIMA. But she needed to build these bridges to make progress in measuring and ensuring water resource health in the Sula Valley. In fact, Rojas drew on a number of skills to bring people together. Working many long hours, she began to think about allies within DIMA who would be interested in coming together under Water Resources. Once she had identified those allies, she worked with them to propose the new unit to the General Manager at DIMA, who eventually approved the idea. Upon her return from Europe, she also moved up the ranks of DIMA's bureaucracy from Technical Assistant to Manager of Water Resources Department with authority over the Watershed, Wastewater Discharge Control, Water Quality Control, and Hydrogeology units, most of these her own creations.

The group began by doing water tests, gathering data to take to decision-makers. They created an "inventory" of the industries emitting effluents at the time. Fortuitously, at the same time, a large company was contracted as part of the sewage evaluation plan to do water testing. Rojas notes that the large-scale testing that company was able to complete showed major pollution of waterways: "they make a characterization of this river up to the point it discharges to the sea…and they found out that especially during the dry season there are about twenty kilometers in this river that, where the dissolved oxygen is zero. So there is no life at all because of the amount of effluents that were discharging at that time." While scientific testing made water pollution visible, the new Department was now able to regulate effluents into the river. Rojas's organizational work had paid off, for it built a bridge between science and policy enforcement. According to Meza, Rojas's years at DIMA "were the epoch of maximum splendor at DIMA…during the time it [DIMA] received numerous awards and recognitions among environmental organizations. Its positive example in the decentralization of water management transcended geographical boundaries" (Meza, R., 2005, p. 86).

Due to this important work, Rojas and other colleagues from DIMA were called in by the Ministry of the Environment (created in 1993) to help draft legislation to regulate water resources. Rojas, in other words, was able to return to an organization in a country where there was little awareness of water and sanitation approaches, re-arrange a government organization to address a new way of seeing water, and then affect the country's environmental legislation in the process. When pressed to reflect more on how she accomplished these extraordinary tasks, Rojas says this:

> I think that the first thing is that people need to believe in what they are doing. Yes. And maybe to have a commitment with the country, with the people you work for. I think that's very clear. If you are not really committed and you don't believe in what you are doing, it's very difficult to really reach those goals.

But commitment by itself is not enough. As an engineer, Rojas had to overcome a view of water as a physical object of engineering analysis (see Figure 7.1 above) and learn to see it as a resource to be protected (see Figure 7.2 above) through new organizational configurations and legislation.

Exercise 50 *Describe Elena's strategy for making water a "resource," visible within her organization. Why does she do things the way she does? How would you characterize her approach? Is it successful?*

7.5   DOING "BIG" DEVELOPMENT

Rojas was proud of the work she completed as a technician at DIMA. By 1995, however, she was ready for a career change. She was offered a position in the Honduran government, one that would allow her to work on the Bay Islands project. Rojas was delighted to be earning three times her former salary, but would soon discover that her new work had in store a host of unforeseen challenges.

In 1996, the government had received a USD$19 million grant from the IDB to institute an environmental program for the entire Honduran archipelago on the Mexican gulf side (see Figure 7.3). The Bay Islands are located in the second largest coral reef in the world, and by the late

1990s, the area had begun to attract many tourists interested in its world class scuba diving and gorgeous ocean setting. Rojas was appointed coordinator for the environmental sanitation sector of the area, which meant she was in charge of the water and sanitation infrastructure that was to be built. In particular, she would be in charge of improving water supply systems and implementing sewage and solid waste systems for the local inhabitants of the Islands. With two other team members—one charged with natural resources management and the other with capacity building—she moved to the islands. The group knew that the coral reefs were already at risk from unmanaged pollution, and she hoped to address this through their efforts.

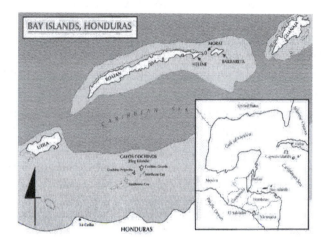

Figure 7.3: Bay Islands, Honduras.
(Source: `http://www.worldstatesmen.org/bay_islands.jpg`. Permission pending).

Rojas had first visited the Bay Islands in 1992, upon her return from Europe. At that time, tourism had not yet flourished the way it would later. There were no more than 20,000 people living on the main island where Rojas and her team would later live during the IDB project. The main island itself is approximately 150 square kilometers in size—a relatively small island. As Rojas described,

> But none of the settlements had any single sewage system. Few houses had latrines, so everything, all effluents, everything was going directly to the sea. And there was not enough water supply, the water supply systems were very poor. So […] the government's idea was to improve water and sanitation coverage because they were aware that tourism will not grow if those public services did not improve. So that's why we were sent to initiate the program.

According to Stefan Gossling,

the IDB project did not address directly the social and environmental problems emanating from escalating human populations. Despite the participatory rhetoric written into the project summary, interviews with local people in the communities of Sandy Bay, West End, and Flower Bay during the summer of 1995 [during the feasibility study prior to the loan] revealed that few residents (apart from a small group of wealthier business owners) were well informed about the project or were consulted in anyway (Gossling, S., 2003, p. 135).

But Rojas quickly set out to involve the islanders. She learned that the locals, though officially citizens of Honduras, had a more complicated sense of their own identities. They saw themselves in relationship to their history as a British colony. Disputed among Britain, Spain, and Holland for more than a century, the Islands finally came under British colonial control in 1643 and incorporated as part of Honduras sovereign territory in 1872. Rojas frames this history as follows:

> The islanders, the people who live in the islands, are very complex because on one side they believe they are British in origin but on the other side, we have a very large population of [Garifunas]. It's a very mixed culture, ethnic composition. Garifunas are black people who used to be slaves two hundred years ago, and just by accident they arrived to those islands. So they have been living there for over two hundred years.

That meant that they were often wary of outsiders, even if those "outsiders" were Hondurans. This knowledge would inform many of the steps Rojas would take in implementing the water and sanitation projects on the islands. She seemed to intuitively understand that if the locals didn't support the new systems, they wouldn't be successful in the long run:

> I think that was a kind of beginning of my social relationship to people who at the end are the beneficiaries of a solution for a public service. Because at that time I was struggling with them, they didn't understand why we were sent to the islands to work for, to do what we were supposed to be doing and we knew we needed to work together with them; otherwise any solution that could be implemented was not going to be sustainable.

But she did not know how to talk to and listen to the Islanders, so she and her colleagues recommended to the minister that a process of "socialization" be undertaken. Working with a number of local NGO's (non-governmental organizations that were frequently staffed by islanders), Rojas's team began to develop a number of workshops with the islanders, hoping to educate and interest them in the sanitation and water projects (see Figure 7.4 for one example of local knowledge sharing across cultures). Rojas believes that the locals saw that tourism was going to happen, and that they knew it could potentially increase their income, so in that sense, they welcomed the growth of the industry.

At the same time, acceptance of the new systems did not happen automatically, if at all. Although Rojas feels the NGOs accomplished more than she ever could, given their close understandings of and ties to the community, she wasn't sure of their long-term effectiveness. When

Figure 7.4: Ana Lucy Bengochea, a member of a Garifuna community who survived a number of hurricanes in the Caribbean, shares her experiences and lessons learned with women who survived the tsunami in Tamil Nadu, India. "Garifuna in Honduras have gained insights and implemented ideas from elsewhere; for instance, hurricane resistant reconstruction techniques from a women's group in Jamaica." (Source: http://proxied.changemakers.net/journal/300510/displaydis.cfm-ID=29 Credit: Groots).

she returned to the Bay Islands almost four years after leaving the project, she "realized that even though all the effort that was done to socialize the program, still people were complaining they didn't like the solutions. Many people were not using them; they thought they were polluting more than discharging directly to the sea the effluents." The islanders had been accustomed to their practice of using latrines that discharged directly into the sea, or into open fields. They were not familiar with or approving of the new systems. Rojas notes, "So it was very complicated. Up to now, I know that finally the water system infrastructure was implemented in the three main urban areas. But still the local municipality doesn't want to take over to start operating these systems so the [central] government has a unit, a technical unit that still is doing the job for them. It's very difficult."

Rojas only stayed on the Bay Islands project for a year and a half. She became quickly disillusioned with how the politicians involved in the project treated the "technicians"—the engineers and scientists. In particular, an election occurred in the middle of the project, and the official who won public office in the election wanted to appoint followers from his own party to projects. Rojas was frustrated by this politicization of her work:

[T]he first thing he [the new minister] did, he started to investigating which people was belong to his party and which did not. So he found out that the team I was working with, we had about two or three people that belonged to the other party. But he didn't really know the technicians, if we belong to one of the other party, and I was happy of that because I was not contracted because I belonged to any country, or to any party; I was contracted because they thought I had the ability to do the job.

By the time of the election, Rojas had been promoted to serve as main coordinator for the entire Bay Islands project, and she was looking forward to making some progress after many delays. Unfortunately, the new minister began to replace her team members with people from his own party, thus slowing the hard-earned forward momentum of the project. Rojas protested to the minister, in writing, asking to be informed of such changes in advance. The minister responded by reminding her that she was "only a technician," and that he would "make the decisions." At that point, Rojas decided to leave the project. Although she had learned to see water in a new way, Rojas did not have the leverage to reorganize the relationship between local and central government water agencies nor the skills to listen to the Islanders. In the Islands, she could not build the bridges between water science, policy, and community. (For a well documented analysis of the politics behind this development project, including how the government politicized the hiring and firing of technical experts, see Gossling, S., 2003.)

Exercise 51 *Build a timeline for Elena's work up to now and to try to identify 1) the challenges she has faced and 2) the strategies she used to counter those challenges. Could you identify two to three principles from other chapters of the book that might explain her strategies?*

7.6   MOVING TO SUSTAINABLE COMMUNITY DEVELOPMENT

The next few years would bring substantial changes for Rojas: she married, had a son, and began a lucrative consulting business. She also began working part-time for an American NGO doing work in Honduras called Clean Water International (CWI). CWI's mission is to help people in developing countries improve their quality of life by supporting the development of locally sustainable drinking water resources, sanitation facilities and health, and hygiene education programs. Rojas would eventually begin working for them full-time, in 2000; at that time, the organization was very small, with an annual budget of $30,000. Eight years later, that amount had increased more than tenfold.

Rojas believes that her role as country coordinator for CWI Honduras is to provide sustainable solutions to water and sewage problems. Sustainability has multiple components; first and foremost, the technical solutions CWI proposes must be accepted by the people. The systems must also be technically *and* socially sustainable, meaning they are designed to last, and to be supervised and maintained by the local population. Rojas and CWI also believe that water projects need to be environmentally sustainable, and that communities should have the legal resources they need to protect their watersheds. Finally, Rojas and CWI have determined that health and hygiene education is a crucial element to sustainably designed projects. But above all in this process, Rojas values the interactions she and CWI have with the communities in which they work. Without this, the water projects would undoubtedly fail:

> Just by having the challenge and learning everyday that if I don't talk to people, if I don't come to people, and if I don't convince [them] of what they need to do in order to maintain and operate their system, we are not going to succeed. They are not going to succeed and we are not going to succeed. [...] Because first of all you start understanding the connection that you can be a very good technical engineer and do your technical projects, your water projects in a very neat way, and you can implement them, that's not really a challenge at all. That's easy to do somehow, you only need to assure the resources, the economic resource. But that challenge [can be stated like this:] once those projects are implemented, what is the key issue to make sure that they will last the time you have planned they should last? So that's something that you as a technical person cannot solve if you do not take into account the people that are going to be taking care of or using those systems.

With this goal as its backbone, Rojas has developed, in conjunction with CWI, a method for community development that takes these sustainability considerations into account. The method is called *community mapping*.

Rojas' view of water as a resource that needs to be protected from its origins to its after-use shapes the way she maps social interactions among those with who she works. She clearly understands that water lives in a watershed, not just in tanks and pipes; hence her need to include all stakeholders, local, and regional, who depend on that watershed:

> We are really working very hard to teach the communities that our water system is not composed only by the infrastructure that is built for the water to come from the mountain to the house, but that the watershed is one of the main components of a water system...we have a challenge to make sure that every water system we contribute to build in the country, it includes the watershed protection. The community should protect the land where the water source is located, either the community or the local municipality. That's something that we facilitate to be done. And then we have a legal process in the country where watersheds, even small areas, could be declared as water production areas, legally. So our goal is to help communities and municipalities to reach that, to get that legalization.

To begin community mapping, Rojas begins by holding a series of one-day workshops with local water "committees" and municipal representatives. Members of local water committees are essentially volunteers from the local communities that have a vested interest in water while municipal representatives are officials at the county (municipio) level. Both are contained within a watershed. Local volunteers are typically "natural leaders," says Rojas, and as a result are typically well respected in their communities. At those workshops, CWI begins with a very simple question about water quality: How many communities have operational systems? How many existing systems need rehabilitation?

What is the actual coverage in each area? What can be done to increase coverage sustainably? She then encourages the water committees to develop and articulate their concerns. The workshops also serve a strategic function in that they raise the visibility of CWI in the region so that they can begin building trust with locals.

Then, using the basic information gathered at the workshops, Rojas and CWI begin to gather specific information about local communities. The organization interviews health workers and school teachers about the water and sanitation infrastructures at the schools, and also interviews members of the water committees about water usage and sanitation habits in the community. They verify the community location using GPS, and they take and analyze samples of each single water source being used by the community. Using this information, the organization begins to build a database containing information about water and sanitation in multiple communities; this information can be used to prioritize projects across the region and is an integral part of the community mapping process (see also Figure 7.5). Rojas describes the process as follows:

> We know how many communities in [a given municipality] do not have at all any water or sanitation system. We know which committees […] need some training in different areas, we know right now which communities are paying a tariff, which are paying a very low tariff. So we got very key information for us to be able to plan what we're going to do over the next four years. And of course, that is a kind of planning tool because we are not going to develop any project because of the mayor, because the mayor tell us, "Oh we need this project here." We will do it according to the priorities we have identified together with the water committees. And for instance, right now when I go back, I will have another workshop seminar with the water committees again to plan for the new projects next year.

Another step in organizing communities around water issues is to create two additional committees in each municipality; in addition to the water committee, there is a watershed committee and a basic sanitation committee, which often involves school children in educating the community about hygiene and health. Through these committees, Elena builds bridges between science, water protection, and community.

These bridges allow CWI and partner organizations, NGOs that are familiar with specific geographical locations, to achieve sustainability in their projects. These partner organizations are often able to provide technical expertise and local knowledge as communities organize and move projects forward. For example, after listening to a community, CWI may partner with a local NGO who works with a mason. That mason can go to the community and train four to five people at the community level on latrine building. The mason will work with the community to build a sample latrine, so that everybody knows how the materials should be used. Then those five trainees can go back to the other families in the community and teach them how to build their own latrines.

The importance of going and meeting with the communities when doing this work can not be overstated. Statistics kept by the municipalities may not always be correct, and organizations like CWI must have a way of gathering more accurate information than already exists. While the

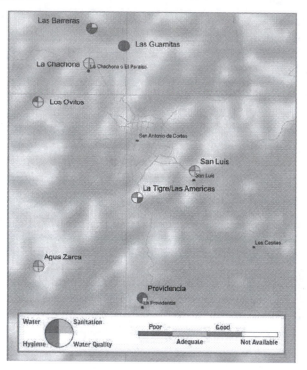

Figure 7.5: Visual results of CWI's community mapping, under Elena's supervision, of Hondura's poorest communities, showing coverage of water, sanitation, hygiene and water quality.

government does have some statistics, they are frequently inaccurate and outdated. In collaboration with communities, CWI may have a target to develop three projects in a particular municipality, where communities have planned to build 124 latrines. "When we performed the mapping exercise," says Rojas, "to identify how many latrines we needed to budget for this year, we went to the municipal statistics, to the governmental statistics."

> We even had meetings with the water committees and we gathered information from different sources and ended up that for those communities we needed to implement 125 latrines. When we did the mapping exercise, that means we went to the community […and] did a kind of assessment in the community, at the end, we realized that it is not only 124 latrines which are going to be needed but close to 180. So that's a very clear example why doing a mapping exercise is very important.

Rojas notes that in her many years of experience, she has seen many projects in Honduras and elsewhere fail because they "never consult the beneficiaries." Rojas does emphasize, however, that her goal is not just listening to and involving communities, but also educating them. Many of the communities she and CWI work with do not understand the connection between poor water and sanitation infrastructure and the high rate of child deaths due to diarrhea and other water-borne illnesses. Instead of viewing communities as passive recipients of "help," Elena and CWI empower community members, including high school students, to become "hygiene educators" in their own communities. In doing so, Elena facilitates communities to become aware of their own problems and formulate their own solutions.

The focus of CWI's education initiatives is to get more schoolchildren involved in educating communities about health and hygiene. Working with the ministry of education, CWI hopes to develop a training program for schoolteachers, which will teach students basic concepts of hygiene, such as proper handwashing and latrine maintenance. "I believe that children are key people at the community level; they will make the changes in the coming years if we educate them," says Rojas. But building infrastructure is a key component here: As Rojas notes, "How will you be able to teach a child, a young child to use the latrine and to keep it clean and why is it important to use the latrine if the school does not have a latrine?" She also believes that education is a key element to the success of any project, and that it can often be the most time-consuming and painstaking part of the project, because change is often slow to happen: "You cannot see from night to day a community with latrines and making sure that they are using them, it is a long process."

In conjunction with the local workshops and committee building, CWI has also developed a strategic plan with the municipalities, focused on capacity building and containing the following points:

- First, over the next five years, each municipality has agreed to create a water and sanitation unit that will eventually be able to advise communities on technical, administrative, and legal issues.

- Second, CWI will organize a series of training seminars for local water communities and for the municipal association of water boards. These seminars are geared toward enabling these groups to better manage the services they provide to their own communities.

- Third, CWI wants to create strategic alliances with government bodies, such as those managing forestry; non-governmental organizations, such as those involved with national water issues; and educational institutions, such as universities conducting research in water quality and training engineers to solve water-related problems.

Rojas hopes that these alliances will help CWI to better map the country, to involve community in mapping and addressing the problems, and to prioritize projects: "Knowing what is the current situation with water and sanitation in the different communities we are targeting will help us to prioritize where we need to intervene and who the partners are and how we should do it with [the] contribution of the stake holders."

Rojas believes she and CWI can also play a role in making sure that future engineers in Honduras are trained to talk to the people. She believes that university education is changing in this regard: "As I told you, we never were trained at the university level about the importance of getting this kind of social relationship. […] This is a real, very new field for them because we [engineers] are not trained in social issues."

A part of being trained in "social issues" is developing a method for assessing the success of projects in the long-term. Rojas and CWI are performing just such assessments.

> And what we do with that monitoring exercise is that we go back to those projects we executed during the last 8-10 years, and we have some specific—a kind of questionnaire again and we monitor how the projects are working. And that's a very important tool that was in fact developed by Clean Water International that allows us in our countries to see what has worked well, what has not worked very well, and how we can improve our new projects, or our current program. And it's also very important for donors here in the U.S. to know if it has been really worth while, their contributions to the countries.

In reflecting on the successes of CWI's projects in Honduras, Rojas says that she feels these projects create opportunities for her fellow Hondurans:

> If they had a chance to get better education, they would be aware the opportunities are there and that they can develop. And I think by doing this kind of water and sanitation projects, it also has an additional benefit. When they build their own water project, they realize they can do a lot more for their own community, that's something they also learn along with the execution of this type of projects. After they [many communities] do their water project in the way we work with them, they know they can go for electricity, or that they can go for having a school at the community. And that's something that also contributes to their own development which is also good.

In sum, Rojas has accomplished significant work by linking science, policy, engineering and communities in a way that allows communities to take responsibility for the future of their water use, management, and protection. Furthermore, she has influenced CWI to adopt this approach in other parts of the world. She has now been promoted to CWI's region coordinator for Central America.

Exercise 52 *By now you have probably read and learned quite a bit about engaging and empowering communities (especially if you read the rest of this book). What do you think of Elena's practice of community mapping? Is the community effectively engaged and empowered? What would you have done differently? Do you know of any other engineers and/or organizations such as CWI that deploy practices for community engagement and empowerment? If so, how are these practices similar or different to Elena's community mapping?*

7.7   QUESTIONS FOR REFLECTION

1. Describe Elena's strategy for making water a "resource." Why does she do things the way she does? How would you characterize her approach? Is it successful?

2. How would you describe the Garifunas? How would you describe the members of Elena's team at the Bay Islands? How would you describe the interactions between the two?

3. How would you describe her transition from large development projects, like the one at the Bay Islands, to community development projects, like the ones she currently works on with CWI?

4. Who are the experts in either type of project? What makes them the experts? What are the different types of expertise at work in SCD contexts?

5. Throughout the different stages of her career, how effective is Elena's disciplinary engineering knowledge (e.g., civil engineering) vs. her other knowledges and abilities?

6. Often in engineering we tend to see water as a physical object to be moved from point 1 to point 2 or as a resource to be managed by systems of water supply and wastewater management. Yet water is also a concept, a symbol, and a social construct that plays a role in many parts of society. As you read this case study, identify and briefly describe the different places in society (e.g., educational institutions, government bureaucracies, local communities, etc.) where different people try to define water in different ways.

7. Once you have identified the places in society where water is defined differently, for each place identify different stakeholders in these places. Who are they? How does each stakeholder see water? What are some of the conflicts and struggles that arise from different stakeholders defining water differently?

8. What strategies did Elena Rojas use to navigate and negotiate among different stakeholders and their views on water? Did her strategies change according to the place in society where she was operating?

9. What did you learn from this case study that might shape your future actions as an engineer?

REFERENCES

Gossling, S. (2003). *Tourism and Development in Tropical Islands* Northampton, MA, Edward Elgar Publishing. 175, 177

Meza, R. (2005). "Municipalización de las cuencas productoras de agua en San Pedro Sula, Honduras." *Recursos Naturales y Ambiente* **43**:80–89. 173

CHAPTER 8

Students' Perspectives on ESCD: A Course Model

"Currently I am involved in developing a school district in Uganda. We are trying to restructure the curriculum and also to provide a way to get water to the school…I try to look at how things can be done in a better and in timely fashion and how everyone can benefit from development."

–Mia[1], engineering student in ESCD class, responding to self-assessment of her relationship to development during the **first day** of class.

"After this class, I believe that my relationship to development as a citizen has changed or …taken another role. I think I should focus on 'developing' what is around me. There is not a single path that can be taken to solve all of the world's problems, and a problem to someone might not even be a problem to someone else even if they are in the same community. LISTEN and be very observant to what is going on around you."

–Mia, responding to self-assessment of her relationship to development during the **last day** of class.

The statements above were made by a student who took our seminar, Engineering and Sustainable Community Development (ESCD). From our point of view, such statements illustrate the importance of coursework that prepares engineering students to think critically about their work with communities. The first quote by Mia reflects an attitude held by an increasing number of engineering students. Their relationship to any form of development boils down to a desire to help people in need through the application of engineering knowledge, even in areas such as curriculum development, which have not been the purview of engineering applications. We encountered many engineering students treating this desire unproblematically; as a result, we set out to help them critically situate their desires and engineering knowledge historically, politically, and culturally in relation to development. As far as they were concerned, before taking this class, as Mia's first quote reflects, they simply "look at how things can be done better and in timely fashion" believing that "everyone can benefit from development." Joe, one of Mia's classmates, made a similar written declaration on his first day of class: "I find my relationship with respect to development as one of having knowledge and being in the position to help others. I feel that it is my responsibility to help others in need and to use my education to improve the world."

[1]All names used in this chapter are pseudonyms to protect students' privacy and confidentiality.

After taking the course Mia re-assessed her relationship to development, questioning her desire to help people in need in far away places. That is, she indicated that she would instead focus on her immediate surroundings and question engineering problem solving. She has come to see that perhaps what appears to be a problem to an engineer is not a problem for a community. She also came to value the importance of listening to those one is supposed to serve. Similarly, Sophie, another of Mia's classmates, wrote at the end of the ESCD course:

> Before taking this class, I believed that sustainable development was nothing but a problem that had a scientific and economic solution. I now believe that in order to achieve any kind of sustainable development, there must be a focus on building relationships and trust with affected communities who must be allowed to contribute their local knowledge and participate in the creation and implementation of solutions.

As their end-of-course quotes reflect, both Mia and Sophie had come to realize that building relationships and trust take priority over any desire to apply knowledge and uncritically help people in need.

What kind of curricular journey can help students to change their beliefs and attitudes towards development in this manner? How could engineering students learn to position and assess their own knowledge and question their desires to help, while finding value in building relationships and learning from local knowledges? This chapter is about student attitudes towards working with communities while taking our course *Engineering and Sustainable Community Development*. More than expecting you to find a similar course on your campus or to make the same realizations as the students in this chapter, we hope to elicit critical reflection on how students like you can begin to change their position towards engineering and sustainable community development (SCD).

8.1   WHAT WAS THIS COURSE ABOUT?

After one year of research and preparation, a team of faculty hailing from the liberal arts, engineering, and environmental science advanced a course for upper-division undergraduate and graduate students entitled *Engineering and Sustainable Community Development (ESCD)*. In terms of learning objectives, by the end of the course, we expected students to be able to

1. identify events, institutions, and actors in the history and politics of development as related to SCD and engineering

2. identify, relate, and describe the role that engineering might play in the different aspects of sustainability: economic, environmental, ethical, and socio-cultural

3. evaluate the strength and limitations of Engineering Problem Solving (EPS) and at least one engineering design methodology with respect to working with communities

4. analyze and evaluate project-based case studies in SCD and select criteria for such evaluations

5. provide and critically assess definitions of SCD and their relationships with engineering

We believe that it is important for you to know the history behind each one of these course objectives so you can gain a deeper understanding of how faculty struggled in developing and delivering each objective and how students wrestled with these and responded along the way.

Although we did not include every student's reaction in our narrative below, we chose to follow two students in their intellectual growth for they are representative of students' profiles and attitudes towards SCD. Karen, a chemical engineering student in her junior year, represents the kind of engineering student that is enduring perhaps the most difficult year in the engineering curriculum and comes to courses such as ESCD with both the weight of curricular demands and the certainty that engineering knowledge and methods can solve most human problems. Daniel, a first year master's engineering student, had begun to question the appropriateness of engineering problem solving and design methods after a brief employment experience as a design engineer. Let's follow their journey through this class.

8.2 "DEVELOPMENT PROJECTS INVOLVE HISTORY AND POLITICS"

How did students deal with the challenge of identifying events, institutions, and actors in the history and politics of development as related to SCD and engineering? First, let's look at how faculty became interested in the history and politics of development to the point that we made it the first of a short list of course objectives. This is not a trivial question with an easy answer. As we have seen in Chapter 2, engineers have a long and complex history in their relationship to development. But we did not arrive at this realization on our own. Our pathway to that history developed out of a struggle with ourselves and our students in trying to understand how engineers relate to humanitarianism.

Our journey to the history and politics of development actually began in a previous curricular experiment in what is called at our university "Humanitarian Engineering." After receiving a large grant from a private foundation to create a program that would change the way we traditionally teach engineering to students, the faculty involved with the grant chose to create an initiative called "Humanitarian Engineering" (HE). Slightly more aware and suspicious of the term "humanitarian," other faculty involved in this grant began a historical and philosophical exploration of the term. We found out how humanitarians emerged in the 19th century as medics and relief workers, became organized under large organizations like the International Red Cross, and played significant roles in World War II, but until the 1960s included no major involvement from the engineering profession. In short, the history of humanitarianism and the histories of engineering for most of the 19th and 20th centuries were not connected at all.

Only when researching the 1960s did we find the work of Fred Cuny, an engineer turned humanitarian, who began rethinking humanitarianism by developing new methods and a new mindset. But Cuny did not represent the general attitude of the engineering profession towards humanitarian practices. Perhaps an anomaly at the time, Cuny was an individual committed to using his engineering and other knowledge to alleviate suffering in humanitarian disasters. Although portrayed as a moral exemplar by some in engineering ethics, Cuny's work is not what our students were doing in their humanitarian engineering projects. Our students were certainly not going to disaster areas after a hurricane or an earthquake to work with displaced and injured refugees.

In this historical journey, we also came across Doctors Without Borders (MSF), perhaps the oldest and most comprehensive approach to humanitarian work by a profession. It became clear that the very recent Engineers Without Borders (EWB) found inspiration in MSF, yet EWB was doing something very different. In short, most engineers that we work with wanted the label "humanitarian" but were doing something else: *community development*. Clearly, our students needed to understand the connections and disconnections of the histories of humanitarianism, engineering, and development.

So if our students are going to be doing community development, we owed it to them and to ourselves to understand the history of how engineers came to be involved in community development in the first place. At that point, we made a thematic shift in our curriculum development from humanitarian engineering to community development.

In our new journey exploring the history of international development, we found that recent histories place engineers, as agents of development, right in the midst of complex geopolitics and ideologies. Clearly, engineering faculty and students, whether they wanted to be called "humanitarian" engineers or "community development" engineers, could not remain outside of history and its politics. Hence, we decided that our students needed to wrestle with the history and politics of international development, including the following key points (most of these are elaborated in detail in Chapter 2):

- The history of international development has been deeply shaped by the ideologies of progress and modernization (Rist, G., 2004);

- In the last five decades, international development has changed emphasis from rapid modernization to basic human needs to structural adjustment to sustainable development (Scott, J., 1998; Rist, G., 2004);

- Being deeply intertwined with Cold War politics, international development had different manifestations in the way China, the USSR, and the US dealt with other countries (Adas, M., 2006);

- International development has continued to reinforce economic differences between countries that were set in place during colonialism (Escobar, A., 1995);

- International development is a contested idea, minimally among those who see it as a moral imperative that "we" must not fail to solve (Sachs, J., 2005), those who challenge us to reconsider its assumptions and approaches (Easterly, W., 2006), and those who invite us to abandon the idea completely (Esteva, G., 1992);

- Engineers, and other agents of international development, might better serve the recipients of development if they move away from being "planners" to being "searchers" able and willing to listen to the people they serve (Easterly, W., 2006; see also Chapters 4 and 5).

Key Terms

Planners: Development workers who have good intentions and high expectations, follow blueprints and plans, unwilling to accept responsibility for outcomes, and lacking knowledge of communities. One who "thinks of poverty as a technical engineering problem that his answers will solve."

Searchers: Development workers that find solutions that work, accept responsibility for their actions, adapt to local communities. One who knows that "poverty is a complicated tangle of political, social, historical, institutional, and technological factors" (Easterly, W., 2006, pp. 5–6).

8.2.1   HOW DID STUDENTS RESPOND TO THIS CHALLENGE?

For this class, students had to select a development project of their choice. In Spring 2008, students selected projects such as

- Colorado-Thompson Water Project

- Columbia-Snake Rivers Dams

- Onion crop irrigation in Senegal

- LED lighting for a community in Ecuador

- Water distribution system for a community in Honduras

- Sustainable mining project in Ghana

- Bielsko-Biala (BB) water treatment project in Poland

- Arsenic removal project in Bangladesh

- Agro-forestry project in Lake Victoria

- Narmada Valley dam project in India

   After one month of classes and readings, students began writing to explore the historical and ideological dimensions of such projects. Specifically, students delved into the following questions:

- How did the project come into being? That is, who first conceived it? When? For whose benefit? Where?

- What were the underlying assumptions behind this project? For example, was this project supposed to help a village or country enhance its quality of life? Freedoms? Economic growth?

- How could the historical period in which the project came to life, from conception to implementation, have influenced what the project was (is) for? For example, was it part of the Cold War? If so, from which side and for what purpose?

For most students, this was the first time that they had to position an engineering project in its historical and ideological context. Karen, a chemical engineering student, chose to research the project on LED lighting for a community in Ecuador being done by her peers in a senior engineering design course. The goal of this project was to design and build inexpensive and portable lighting devices for villagers living in a remote part of Ecuador. After briefly exploring the questions above in relationship to this project, Karen concluded that "the cultural ideology [of this project] is in line with current trends in foreign development." Realizing how this exploration was changing her perspective towards development projects, she acknowledged that "a historical look at foreign aid helps new engineers in the program avoid the pitfalls and mistakes of previous projects." She quoted Easterly by writing that

> the current wave of enthusiasm for addressing the world's poverty is doomed to repeat the cycle of its predecessors: idealism, high expectations, disappointing results, cynical backlash if rich countries don't address the tragic history of well-meaning compassion [that] did not bring…results for needy people.

She was rapidly becoming skeptical of development projects yet unwilling to completely qualify her peers' project as a development failure. After all, she was in the process of joining the program on campus where the LED project originated. She wanted to have it both ways: to criticize development, yet excuse her peers project from being part of it. She concludes:

> the [humanitarian engineering] program appears to be striving to chip away at the tragedy [of decades of failed development projects mentioned by Easterly] …I believe this project was a success because the students approached the problem as 'searchers.'

As the course went on, her views would change.

Daniel, a graduate engineering student, selected the Bielsko-Biala (BB) water treatment project built for a Polish town in the mid-1990s. He accurately mapped the political and ideological origins of the project to Poland's shift from communism to market capitalism and to Poland's desire to join the European Union: "these political pressures would have weighed heavily on the minds of all parties involved in the approval of the…water project." Furthermore, he identified a wide range of actors involved in the project, from the World Bank to various public and private entities involved in its construction. Having not yet learned how to assess community participation and empowerment in development projects, Daniel took the assessment of development organizations at face value: "the [BB] water project was a success for the World Bank and for the community of

[BB] in many ways…[the] infrastructure was successfully updated to meet almost all of the promises initially made in the loan agreement." But he still wondered about the impact of the project on the community:

> The World Bank loan agreement did not even mention the effects on the community outside of technical issues." And thus he concluded that "[the World Bank] approach is consistent with a Planner's approach to problem solving rather than working from the ground up.

Powerful revelations about the project's impact on the community came to Daniel later in the semester.

Clearly, these students were learning for the first time to position development projects in their proper historical and ideological contexts amidst many actors and institutions. Yet most of them resisted being critical of development. Karen was writing as an advocate of her project, her peers, and the program she was about to enter. Daniel was not sure how to interpret the World Bank's report in relationship to the community, so he took it at face value. Until students had the opportunity to develop their critique, we could not expect them to do anything but analyze projects in engineering terms, mainly in terms of "efficiency."

8.3   "DEVELOPMENT PROBLEMS ARE MORE THAN JUST TECHNICAL PROBLEMS"

How did students respond to the challenge of identifying, relating, and describing the role that engineering might play in the different aspects of sustainability: economic, environmental, ethical, and socio-cultural? During our exploration of the history of engineers in development, we found that when engineers became interested in sustainability they tended to view it as a technical problem to be approached with technical solutions (see Chapter 2). At the same time, through our study of sustainability and sustainable development, we discovered that these concepts have many complex dimensions that are intertwined with the technical. If our students were going to be involved in SCD, they needed to learn the many dimensions of sustainability beyond the technical. We challenged students to explore questions such as

- How can a project or system be affordable for a community in the long-term without placing its members in huge debt with external creditors so that the community loses its economic self-sufficiency? (Economic dimension)

- Should sustainable development (SD) be about sustained growth or about the protection of natural resources even at the expense of growth? How are the biosphere and the technosphere related? How can engineers design artifacts and systems (technosphere) so as not to place a burden on the biosphere? (Environmental dimension)

- How can we begin to understand what a community is? What are effective ways to facilitate that communities chart their own destiny through community development projects? How can engineers learn to listen to communities? What if much or even most development to date has been detrimental to communities? Where do we go next? (Socio-cultural dimension)

- What if in order to have SCD we need to ensure communities'

    - local economic diversity,

    - self-reliance,

    - reduction in use of energy and careful management,

    - recycling of its waste products,

    - protection and enhancement of biological diversity,

    - careful stewardship of natural resources, and

    - social justice?

8.3.1   HOW DID STUDENTS RESPOND TO THIS CHALLENGE?

Continuing with their chosen development project, students now had to explore these dimensions of sustainability. Specifically, students explored these questions:

- How does my project fare when evaluated against the dimensions above? What dimensions does it meet? Which ones does it lack?

- How does my project fare in terms of placing a burden on the biosphere?

- Are there any dimensions missing that I should consider in my project for a more comprehensive treatment of SCD?

Again for most students, this was the first time that they had to position an engineering project against dimensions of sustainability, especially one that encompasses community capacities. For Karen, this challenge was a significant learning experience. She began to change her advocacy for the LED project and the minor program that housed it. Instead of blindly praising the project, as she had done forcefully earlier in the course, she now evaluated the project against very specific dimensions as follows:

...a successful and appropriate community development project would create more opportunities for local communities to expand their financial capacity. This could be done by increased demand of their products or a broader market from which to purchase, for example. From my observations of the Ecuadorean lighting project, this criterion was not met. These products were made entirely from imported materials [which] limited the diversity of the local economic market...it appears as if no plan or method was ever designed to deal with the waste created when the lighting units die out...In conclusion, this project failed on most accounts to meet the dimensions and criteria by Bridger, Luloff, McDonough and Braungart [established in class], and myself. Yet, the project was hailed as a success by the faculty advisor and the senior students. At this point, I am unable to say if this contradiction is the fault of the criteria proposed or the self evaluation of the engineers involved.

Karen had acquired a healthy critical distance from the project even though she still wanted to join the Humanitarian Engineering minor program which would require her to be in one of these projects. We noticed that community grassroots activist Gustavo Esteva, from Oaxaca, Mexico— through his writings and class visits—had perhaps the single most significant influence on our students' thinking. We think this was because students saw him as "representative" of indigenous communities (in ways that they can never see us) and/or because of his eloquence in making them question their desire to help. In any case, Karen, like most of our students, was deeply influenced by Esteva and came to question the desire to help of her peers and its impact on the indigenous communities:

> It is plausible that [indigenous people] were indignant, excited, or indifferent to the lighting units and their local impact. For example, some of the indigenous people might have been offended by the notion of needing "help" to obtain these lighting units. As I learned in class from Gustavo Esteva, help is not always the most beneficial thing and not welcomed by many underdeveloped nations.

Using the dimensions above, Daniel also learned to read World Bank reports differently with a healthy dose of skepticism. Even when relying mainly on just one report on the BB project, he now knew to look for:

> The BB Water Project did not significantly aid local econom[ic] diversity. Construction and design of a major engineering project requires the efforts of hundreds of people with diverse skills. Unfortunately for the citizens of the BB area, most of these inputs came from outside of Poland. Welsh engineers, not Polish, were in charge of technical design….

Daniel also learned to assess what the project could have done for the community but did not, such as questioning traditional water treatment techniques:

> The BB project missed a golden opportunity to promote self-reliance. Instead of promoting alternative water systems (i.e., methane producing centralized plants, ecotoilets, etc.) that may have been viable replacements for the failing system, the WB provided funds for 'modernization' of the existing water treatment facilities…. Central wastewater treatment requires energy to operate…[where] sewage is seen as waste rather than as a resource stream…[and where] recycling of resources is made difficult if not impossible….

Perhaps more importantly, Daniel realized that no new knowledge or alternatives could emerge without community participation:

Without local involvement (or local initiative) and careful study of the society and environment, a proper alternative to the BB treatment facility cannot be chosen…the crux of the problem is that there was no discussion of alternative approaches by the parties involved in the BB Water Project. The WB, NFEP, and AQUA chose western (in this case Welsh) water treatment technology as superior….

Both Karen and Daniel had taken important steps towards understanding engineering projects beyond their technical dimensions. Two months into the course, they had come to realize that development projects, large and small, also entail economic, environmental, and socio-cultural dimensions.

8.4 "ENGINEERING PROBLEM SOLVING AND DESIGN METHODS HAVE STRENGTHS AND LIMITATIONS WHEN APPLIED TO DEVELOPMENT PROJECTS"

How did students evaluate the strength and limitations of Engineering Problem Solving (EPS) and at least one engineering design methodology with respect to working with communities? We wanted students to have a sense of how engineers have engaged communities throughout the short history of international development. Fortunately, we found a compelling case study of engineers who in the 1960s created an organization—Volunteers in Technical Assistance (VITA)—with the goal of sharing technical knowledge with communities outside the US. Working from within the military-industrial complex, VITA engineers created technical manuals with the hope of *transferring* technical knowledge to communities that, according to the engineers, needed it (Williamson, B., 2007). Also, our students read another comprehensive account of engineers, some young as them, working in the big picture of development within large development organizations (Jackson, J., 2005). Furthermore, students worked through the two case studies presented in this book (Chapters 6 and 7). These depictions of engineers working in concrete projects had a particularly powerful sway over students' views, perhaps more so than theoretical arguments about development. As students read these historical and ethnographic depictions of engineers in development, large and small, they begin to realize the strengths and limitations of engineering problem solving (EPS) and design when they are applied to community development cases. In class, we dissected together the EPS methodology that students learn in most engineering science courses and began assessing EPS against key approaches to community engagement such as listening (see Chapter 5). Finally, we offered students an alternative approach to understand, analyze and value different perspectives, including those within a community: Problem Definition and Solution (PDS) (see Chapter 5 for a full description).

8.4.1 HOW DID STUDENTS RESPOND TO THIS CHALLENGE?

Continuing with their chosen development project, students now had to imagine how their project could have been otherwise by applying the Problem Definition and Solution (PDS) approach outlined in Chapter 5. They had to imagine being part of the following group of stakeholders who at the beginning of the project actually defined the problem differently and provided different alternative solutions:

1. An engineering graduate from our university who took her fair share of engineering science courses did a humanitarian project in senior design but DID NOT take our ESCD course

2. A VITA engineer *OR* an engineer working for a big development organization

3. Rani Natarjan, the engineer from the Sika Dhari's Windmill project, *OR* Elena Rojas, the engineer from the Community Mapping case study in Honduras (see Chapters 6 and 7)

4. At least two (2) different perspectives from the local community that is supposed to benefit from this project (by now students knew that a community had many voices)

5. The student him- or herself, who after taking this ESCD course had knowledge of criteria for SCD, key considerations about community, and now understood the strength and limitations of EPS and design approaches to development

6. At least two (2) more key and relevant stakeholders (not from the local community) whose perspectives will significantly shape the project and be shaped by the project

Students had to map the perspective of each stakeholder, imagine how each perspective would have defined the problem and proposed a solution differently, mediate (and perhaps reconcile) among competing perspectives, and consider adjusting their own perspectives in order to move the project forward (or to stop it altogether).

Discovering one's own biases. After describing in detail each stakeholder's perspective in the LED flashlight project, Karen made significant realizations about individual stakeholders. She clearly recognized that the engineers coming from her university would have a limited perspective on community, especially if they carried out the design on campus far away from the community: "The atmosphere of learning at [my university] is very challenging and requires autonomous [individualistic] participation. In my opinion, this environment makes it difficult for engineers to grasp the concept of community participation and design." She went on to position each stakeholder, including herself, in his or her historical and ideological context, allowing each stakeholder to define the problem differently and propose different solutions. Through this process, she became aware of her own perspective and biases:

> As I was writing this [analysis], I discovered a strong bias in my opinions. I had a desire to make every perspective echo some of my own personal desires. My experiences in the ESCD course gave me a sense of idealism. I had a strong desire to do the project 'right,' no matter the cost, or not do it at all.

Imposed solutions, even if environmentally sustainable, won't work. Drawing from her learning that sustainable development and social justice in a community go hand in hand, Karen realized that even an environmentally friendly alternative to battery usage and disposal should not be imposed for it might be unfair to the community:

It would be cruel of me to expect or project my own ideal standards for others when they have yet to acquire their own values and set their own criteria for, say, battery usage. While I can see and perhaps teach about the damage old and leaky batteries can cause, I cannot fairly ask the indigenous people of Yachana to do without power when I use thousands of kilowatts of electricity….

Keep your own motivations and desires in check. Furthermore, her analysis of her and other students' perspectives made her realize how much EPS and the grading system motivate students' desires to view their projects as successes, even when they are not, with concrete problems and exact answers:

If I were a student on this senior design project [LED flashlights], I would have some strong desires for the project. As a student who is consistently 'graded' on performance I would want this project to be hailed as a 'success' if not a 'great success.' However, just because I am handed a problem does not mean there is a perfect solution…. There is no 'correct answer' for such open ended problems, as I have been trained to expect in engineering. As an engineering student, I will be required to shift my desire for a concrete answer even though it may not exist.

Discovering humility. Finally, and perhaps more importantly, Karen adopted a sense of humility by realizing that her engineering knowledge and technical solutions are not the silver bullet to solve community problems, and that knowledge should be *bilaterally exchanged*:

It would also be unfair of me to expect a foreign community, with whom I have no real contact, to just accept my systems, programs, and technologies as the end all solution to their daily problems…. Instead of trying to project my own reality on other people, I should try to gain knowledge of their cultural values, desires, and needs from me as an engineer.

Shifting your own perspective. Like Karen, Daniel mapped the perspectives of multiple stakeholders to be involved in the BB water project, imagined different problem definitions and solutions from each stakeholder, and assessed the implications of these solutions to each stakeholder. Taking the role of mediator, instead of traditional engineer, made Daniel realize that the biggest challenge in a community development project might be to shift his own perspective. To do this, he needed to learn to "live in community" in order to become an effective translator/mediator who can establish trust among competing perspectives:

Suddenly, the rather routine job of designing some pipe layouts and plugging in off-the-shelf tanks and pumps becomes more challenging. This massive change in scope would require changes in location, knowledge and desires [all these elements of perspective] on my part…. If this project is going to be sustainable it will require my presence and continuing support for a significant amount of time. A project should establish a relationship; it should be more like a marriage than a one-night stand. Any good relationship requires a level of communication that can't be obtained through an interpreter, so I would need to make every effort to become fluent in Polish…. The only way to truly learn a language is to be immersed in it, and the only way to truly understand a community is to live in it, so a change in [my] location is in order.

Questioning your desire to help. Having read accounts of "expat" development engineers who live abroad in neighborhoods sheltered from the daily lives of the communities that they are supposed to serve (Jackson, J., 2005), and after becoming suspicious of their and his own desire to help, Daniel concluded that

I firmly believe though, after taking this class, that it is irresponsible to "help" a community that you don't live in. To be clear, I mean actively living in the community and interacting as a citizen on a daily basis. This is not the kind of "living in community" gained by living in a gated community of expats.

Valuing local knowledge. Finally, Daniel humbly acknowledged that he has much to learn from local knowledge and that his desire to be paid for his services should not come above hiring engineers from the community:

Most of my education [for this project] would come from people without college degrees or any technical background. Turning expectations upside down and becoming a learner will require serious humility on my part and may challenge my desire to be the 'expert.' Moreover…. my desire to earn a living and work on this project needs to be subservient to the well being of the community. Perhaps it would be best to have a local engineer or technician fill the role of technical lead.

Both Karen and Daniel took perhaps the two most important steps that aspiring engineers can take in learning to work with communities. First, they learned the strengths and limitations of their own engineering approaches and methods. Second, they learned to recognize the biases of their own perspectives as outsiders and how much they could actually learn from community knowledge. In short, they began to learn *humility*.

8.5 "I LEARNED HOW TO MEASURE MY DEVELOPMENT PROJECT AGAINST SCD CRITERIA"

How did students analyze and evaluate project-based case studies in SCD and select criteria for such evaluations? As we have seen, students learned to position their projects in relevant historical and ideological contexts, to evaluate the project's different dimensions of sustainability (economic, environmental, and socio-cultural), and to imagine how their project could have been otherwise. In addition, students as a class generated an extensive and comprehensive list of Crucial Questions for Understanding Community Struggles (too extensive to include here). Then they had to choose questions from this extensive list and generate their own specific questions and criteria in order to evaluate the project in relationship to its impact on community. Finally, they had to write a participatory learning plan (PLP) where students had to make recommendations to those who implemented the project on specific strategies to truly engage the members of the community as equal partners in their project.

8.5.1   HOW DID STUDENTS RESPOND TO THIS CHALLENGE?

By now Karen had exchanged her advocacy for the project, and the HE minor program, for a healthier critical attitude that allowed her to ask difficult questions, including

- "Does the project empower people or make them more dependent on outside forces? Outside markets? Outside technologies?

- Is there a long-term plan to maintain the project?

- Is this the right project? Something the community wants? Is it feasible and appropriate in the setting?

- What potential unintended consequences could the project have?

- What are the environmental consequences of the project? How do we assess this?

- How do you determine project goals in advance? What projects does the community support and value?"

As Karen challenged herself to answer these questions from project documentation and interviews with participants, she came to a striking realization. Had students in the project been challenged to consider these questions, Karen writes,

[They] may have even modified their flashlight design to accommodate community desires or needs. However, the students involved in the project were not even asked to consider these aspects in their analysis…. Apart from technical knowledge, the [program] students were not prepared for the other aspects of development work, such as community participation.

After generating his own questions and criteria to evaluate the BB Water Project and its impact on community, Daniel went beyond official project reports to research local documents and books on water and traditional farming in Poland. Sadly, he realized that "the community aspects of the BB Water Project were almost completely ignored by all parties involved." Rural farming communities were particularly ignored: "The BB Water Project was the result of urban elites working with outside help [development banks and agencies] to implement their vision of the future without significant consideration for effects on rural farms." As he learned from class readings, there is a strong connection between sustainability and community empowerment. Relying now on new evidence, Daniel confirmed that by ignoring the rural farming communities the BB water project had also made farming more unsustainable:

Traditional Polish farming [was] largely organic by necessity, since fertilizers are expensive and hard to attain. Irrigation has historically not been a serious option because the vast majority of private farms draw water from private wells with manual pumps. In place of artificial fertilizer and heavy irrigation, Polish farmers have tended small plots of crops suited to local conditions and used crop rotation to keep the soil healthy.... Sadly, it seems there are many plans for Polish farms and the majority involve changing a system that has worked sustainably for hundreds of years... the completed BB Water Project aids the industrialization of traditional farming and damages the way of life of rural Poland.

Humility allowed Karen and Daniel to ask difficult questions and begin shifting their perspectives away from exclusively considering engineering methods and towards a view that incorporated community interests.

8.6   HOW STUDENTS RESIST

The sources of resistance to a course like ESCD could be many. Scheduled as a liberal arts course in a technical university, ESCD could have led some students to think that it was an "easy humanities" or a "walk in the park" course. Yet when they encounter significant intellectual challenges and higher time commitments than expected, some resist and fight back. Although we experience this form of resistance from some students in every course we teach, we are interested here in the resistances that derive from students' early conceptions of and assumptions about engineering, development and community.

Holding on to engineering. As we have seen throughout the book, engineering students come to value engineering problem solving (EPS) as the dominant method to solve problems in their curriculum. After solving hundreds (often thousands) of problems using EPS, engineering students come to highly value EPS and strongly identify with it by their junior and senior years. Our ESCD students were no exception. They came to our class with strong beliefs about what EPS could do to solve problems, including those of communities who they perceived in need of solutions.

The transformations that Karen and Daniel underwent did not come easy as all students came into this course with their own conceptions and assumptions about engineering, which often make them resist a critical analysis of what is soon to become their profession. Dave, a mechanical engineering student, responded, "I feel my relationship with respect to development is one of having knowledge and being in position to help others." Like Mia (at the course outset), Dave boiled down his relationship to development to a desire to help people in need through the application of engineering knowledge. But unlike Mia, who was willing to change her attitude throughout the course, Dave was unwilling to let go of engineering problem solving (EPS) and eventually dropped out of the course. By now, we have come to realize that our course is not for everyone and may not be able to reach all students, particularly those whose identities are deeply entrenched in EPS.

Yet some students who held strong to EPS continued to resist a critical engagement on the appropriateness of engineering for SCD. Some resisted critical assessments of EPS and engineers involvement in development, perhaps because, as discussed in the Introduction, they hold problematic beliefs and attitudes, particularly in the power of technology to transform society and in the universalism of technological applications across cultural borders. Also, most students who took our ESCD course and Senior Engineering Design II simultaneously (having taken Senior Design I the semester before) initially assumed that "design for industry" methods and practices could be extended to "design for community" (see Chapter 3).

Development and me. At the beginning of the course, we asked students to describe their own relationship to development as citizens. Not surprisingly, at this early point in the course students described their relationship to development in terms that were familiar to them. Like students we have encountered elsewhere, this group has learned to view a "citizen" as an individual person who obtains knowledge through formal education, secures employment, pays taxes, and seldom contacts government officials to complain.

TJ, a materials engineering student, viewed his relationship to development as an active taxpayer who might entreat his government to action: "As a citizen the most obvious way we can support community development is by tax dollars. If there are no plans in place to use tax dollars then the citizen can urge government for these programs or elect leaders that will support the projects." Also viewing himself as a taxpayer, but a passive one, Daniel initially said "My relationship to community development as a citizen is mostly passive and financial. Taxes and donations pay for projects that I do not have input or do not take the time to give input." Some of these students resisted the notion that development is a set of practices embedded in larger historical, institutional, and political contexts. Perhaps unaccustomed to seeing human activity in a larger sociological context, these students held onto individualistic notions of development for a while.

Interestingly, two graduate students positioned themselves either as part of Western countries or as members of multinationals. Geert said, "I am a graduate university student in a developed Western country. My understanding of development is quite informed but highly theoretical—I lack firsthand experience." Liz, the second of these students, answered: "Most of my life I've been at the receiving end of development. When I worked for Exxon Mobil, I was on the implementation side of development." Perhaps due to their graduate training in international political economy, these graduate students came into the course with enough awareness to position themselves in the West or inside a multinational corporation and recognized the privilege of such positions. These students had an easier time accepting the wider sociological and political dimensions of development.

Community as a whole. As reflected by Mia's quote at the beginning of this chapter, before the course, most students held conceptions of "community" as a homogeneous group of people in need of students' help in the form of an engineering solution. Hence, many students resisted the idea of critically questioning their own desires to help communities. Also, most students resisted the notion that communities have divergent and often conflicting perspectives within them, making it difficult, if not impossible, to treat them as a single "client" or "customer." As we have seen in Chapter 3, many of these problematic assumptions about community are reinforced by Senior Design courses.

Yet by the end of the course, most changed their conception of community and their relationship to them.

8.7   TRANSFORMATIONS

By the end of the course, students' understanding of the importance of community and listening had expanded significantly. For instance, TJ had changed his relationship to development from one of taxpayer to one requiring responsible inquiry on the impact that a project might have on community: "If a person is to volunteer or support a development project as a citizen, I would say it is their responsibility to find out about the workings of the project and see if it seems to support community development. It would be a bad idea to blindly support any project, because more harm could be done than good."

Daniel had one of the most revealing transformations, from passive taxpayer to active listener and participant:

> As a citizen, I have realized I can't stand on the sidelines and leave the development issues to others. My perspective matters and at the least, I should express my concerns. As an American, our society has imparted an almost parental role upon the attitude of development (a desire to "help"). The world is not always supportive of this view and we must address development critically in light of the world's concerns.... Don't get caught up in strict problem solving techniques. Actively listen to and critically assess different points of view. Avoid apathy. Ask questions.

Perhaps disillusioned with development after her critical analysis of the LED flashlight project, Karen distanced herself from development, maintaining a relationship only in her capacity as US citizen and member of her immediate community: "I don't feel related to development except that I am a citizen of America. The closest I get to development is my community service."

Geert, the graduate student who initially position himself within academia in a Western country, while acknowledging that his position had not changed much, changed his attitude towards helping:

> I came into the course as a development skeptic and remain one. I still have a desire to be involved, but not to *help*. My relationship to development is unchanged. I remain (only) a development academic... [but I need] to be aware of my own (and others') knowledge, location and desire. To watch as well as to listen. To not *help*. And to travel a lot, but also to work more within my own community (underlining in original).

Liz, the second graduate student, frustrated by the lack of practical experiences in the classroom and the hypothetical nature of many of the course's exercises, declared that she wanted to become involved in the actual "doing" of a development practice:

> I have now become a more critical observer of any form of development projects in my community and others... [yet] as an academic interested in what we have studied this

semester it is important to actually participate in projects. So I will, hopefully, get on board with one of those projects be it humanitarian engineering or in my own community to gain another perspective—the 'doers' perspective.

Clearly, after taking the course, students repositioned themselves from the earlier roles as relatively passive individual citizens to engaged critical observers, and in some cases practitioners, unwilling to take development for granted. Our subsequent offering of this course mirrored this self-conversion of students from dispassionate technocrats of development to, by and large, involved and passionate advocates for people and community. Some of the most interesting changes in viewpoint in both groups came from the international students from rich families who, by upbringing, might be expected to side with government bureaucrats, higher-ups, and technocrats, but who reluctantly started appreciating, and empathizing with, the views of the people on the receiving end of development. Most students had become aware of the power dimensions in development, some have come to question their desire to help, and many now realize the importance of respecting, listening and empowering others to participate. In short, community has become visible to these students.

REFERENCES

Adas, M. (2006). *Dominance by Design: Technological Imperatives and America's Civilizing Mission.* Cambridge, Harvard University Press. 188

Easterly, W. (2006). *The white man's burden: why the west's efforts to aid the rest have done so much ill and so little good.* New York, Penguin Books. 188, 189

Escobar, A. (1995). *Encountering Development: The Making and Unmaking of the Third World.* Princeton, Princeton University Press. 188

Esteva, G. (1992). Development. *The Development Dictionary: A Guide to Knowledge as Power.* W. Sachs. London and New York, Zed Books: 6–25. 188

Jackson, J. T. (2005). *The Globalizers: Development Workers in Action.* Baltimore, MD, John Hopkins University Press. 194, 197

Rist, G. (2004). *The History of Development from Western Origins to Global Faith.* London, Zed Books. 188

Sachs, J. (2005). *The End of Poverty: Economic Possibilities for Our Time.* New York, Penguin Press. 188

Scott, J. C. (1998). *Seeing like a state: how certain schemes to improve the human condition have failed.* New Haven, Yale University Press. 188

Williamson, B. (2007). *Small scale technologies for the developing world: volunteers for international technical assistance, 1959-1971.* Society for the History of Technology, Washington, D.C., SHOT. 194

CHAPTER 9

Beyond Engineers and Community: A Path Forward

The main point of this book is that community should be *the* central concern for engineers involved in SCD work. We hope that by now you are convinced that community's needs and desires should be articulated and decided by the community, not by engineers or anyone one else. Also, the problems and solutions associated with those needs and desires should be defined and negotiated primarily by the community. The engineer's role is to facilitate, if invited, and to listen and learn.

If this book has been successful, the beliefs and assumptions that most engineers hold about technology and its role in SCD have become more visible to you in ways that you feel comfortable questioning and transcending, even if it is you who holds these beliefs and assumptions. It is not within the scope of this book to analyze the historical and philosophical roots of these assumptions and beliefs. For this, we will provide you with a suggested list of readings that will help you in such exploration. At this point, we just want to invite you to keep them present with you as you embark in future SCD projects by asking questions such as

- How much am I motivated to carry out this particular SCD project by my belief in the *power of technology to transform this community?* What are the consequences of holding such beliefs paramount over others, such as the right of the community to make its own choices, including not using my proposed technological solution? (revisit Chapter 6 to see how engineers actually did this). Suggested reading: Marx, L., 1987.

- How much am I influenced by the *ideology of modernization*, particularly the belief that a socially engineered order, informed by science and realized through technology, will bring progress to a community? Suggested reading: Scott, J., 1998.

- How much am I operating under the assumption that *technological solutions can be universally transferred and applied?* Suggested reading: Adas, M., 1989.

By now, we hope you have come to appreciate the importance of the *history of development* in shaping current institutions, practices, ideas, and assumptions about how engineers work with community. In Chapter 2, you have seen how every epoch of development has positioned engineers differently with respect to community, often in problematic ways. The present is no different. Clearly, current engineering practices in SCD have been shaped by this history and, in most cases, communities continue to be ignored, disempowered, or simply treated like an industrial client, in the best of cases. Remember this history even as new approaches to engage communities emerge in the future. Suggested reading: For an alternative analysis of development by an engineering educator, see Chapter 4 of Caroline Baillie's *Engineers within a Local and Global Society*.

In Chapter 3, we explored the problems of adopting *design for industry* to *design for community*. As of this writing, UNESCO and Daimler announced the 2009 winners of the Third Mondialogo Engineering Award, a very impressive collection of engineering SCD projects from around the world (see list of winners at http://www.mondialogo.org/). Yet we worry that many of these projects were carried out under similar assumptions of design for industry as the project that won the "Exceptional Student Humanitarian Prize" highlighted in Chapter 3. We invite you to apply the critical lens that you learned in that chapter to these winning projects and ask to what extent they are adopting design for industry assumptions, practices, and processes when designing for community. Suggested readings: Architecture for Humanity, 2006; Cooper-Hewitt National Design Museum, 2007.

In Chapter 4, we made the case for why community should be at the center of engineering for SCD, where the challenges for engineers to engage communities come from, and how to begin preparing for working with communities. We encouraged you to develop strong self-assessments and strategies for testing your assumptions about and commitments to communities. Communities are often heterogeneous, shape-shifting entities: they are rarely static or easily understood. You must reflect on community members' relationships with one another, with place, with power and privilege, and with their purpose(s); you must embrace SCD as a challenge requiring different tools—such as humility, self-awareness, and a willingness to "fail"—from the kinds of engineering work you have been prepared to do in most of your classes or industry experiences. Suggested readings: Kidder, T., 2004; Easterly, W., 2006.

In Chapter 5, you learned that contextual listening is one of the most important competencies for engineers who want to work in SCD (or in any engineering activity for that matter). From exploring large and small development projects, a recurring lesson emerged: that failure to listen to and meaningfully address community perspectives played a significant role in the failure of such projects. Although barriers exist to enacting contextual listening, we encourage you to actively seek opportunities to practice it, so as to bring about its benefits: contextual listening can 1) counter biases, 2) foster a community-centric approach to problem defining and solving, and 3) integrate multiple perspectives and sectors. Suggested readings: Burkey, S., 1993; Slim and Thomson, 1995; Salmen and Kane, 2006.

In Chapter 6, you saw how engineers willing to listen to community ended up building a project entirely different from what they had originally planned, based on particular community members' input. In Chapter 7, you learned how one engineer transitioned from working in large development projects, operating under a highly mechanistic view of water, to working with communities, empowering them to map their own water use and take care of water as a resource. As *community mapping* becomes a more popular approach in community development, we invite you to engage this approach carefully, always keeping the interests of community first, and paying close attention to how mapping space always alters the power that communities have over their own future. Suggested reading: Rambaldi et al., 2006.

In Chapter 8, you saw how students were able to transform their assumptions and attitudes towards community, even within the constraints of a one semester class. Be willing to search for and take courses that will challenge your assumptions and beliefs about engineering and its relationship to communities. Read accounts by engineers like Fred Cuny (Cuny, F., 1983; Cuny and Hill, 1999) or by medical doctors like Paul Farmer (Kidder, T., 2004) who completely questioned their traditional engineering or medical approaches to serve communities facing humanitarian crises. Engage other works on engineers and community development aimed at helping students and faculty understand how to promote more just and sustainable projects (Baillie et al., 2010).

9.1   WHAT MIGHT BE MISSING FROM THIS ACCOUNT OF ENGINEERING AND SUSTAINABLE COMMUNITY DEVELOPMENT?

The focus of this book has been mainly on the relationship between engineers (E) and sustainable community development (SCD). The importance that we have given to community and engineers in all chapters could give readers the impression that these are the main actors that matter in SCD. You might have created a mental picture of SCD that looks like this:

Engineers ↔ Community

Although *community should always be central*, engineers and community are not the only two actors in SCD. The relationship between them is only one among many in the larger context of SCD. There are other stakeholders and relationships that are important for engineers to know, understand and value in SCD. Although it is not within the scope of this book to elaborate on and analyze all elements, stakeholders, and complexities of SCD, it is important to highlight them so as not to mislead you into thinking that engineers and community are the only stakeholders who matter.

Context. As you learned in Chapter 2, historical and political contexts matter. The Cold War served as a context for post World War II international development until 1989. Similarly, there are currently a number of contexts that shape what is and is not possible in SCD, and you need to pay attention to these. *Geopolitics* is one of them. What happens between guest/donor and host/recipient countries shapes the conditions for SCD. For example, given the current troubling relations between the US and countries like North Korea, Iran, or Venezuela, it would be next to impossible for US engineers to initiate SCD projects in those countries.

Internal conflict is another. Even within countries with friendly and stable relationships with the US, there might be internal conflict that would make it extremely difficult to initiate SCD projects.

For example, certain regions of Colombia–a country with which the US and most European countries have excellent relations, making it prime for SCD project funding—are immersed in armed conflict where SCD would be quite dangerous if not impossible.

You should also pay attention to *local governance*. Even within countries that exhibit relative peace throughout their territory, local governance practices might play a determining role in SCD projects. Highly bureaucratized practices, corruption, and nepotism, for example, would make it difficult to obtain permits, data, access to resources, etc.

And don't forget *ideology*. Donna Riley has shown how the ideologies of militarism, colonialism, racism, and sexism have influenced engineering practices (Riley, D., 2007, 2008). More recently, Riley and Niuesma have shown how the ideology of neoliberalism has influenced engineering projects for community development (Nieusma and Riley, 2010). Everywhere you go, you will likely find pervasive ideologies shaping assumptions and constraints for SCD projects.

Institutions. In all, SCD projects and related activities you will encounter a wide array of institutions. This is not the place to launch an analysis of which kind of institution might be more appropriate for specific circumstances. We just want you to be very aware that the kind and size of institutions involved greatly influence the resources available to and the constraints placed upon specific SCD projects:

- Likely, you will find *development banks* that can be large and with lots of political and economic power that would allow them to impose conditions on countries and local governments.

- *Small micro-loan organizations* may hold more equitable and just lending practices, and hence be more conducive of SCD.

- *International organizations*, such as the UN programmes and specialized agencies that make up the UN system, might also be present in the locality where you are planning to develop a SCD project. Their presence, practices, rules of operation, etc., facilitate or constrain what can be done on the ground. For example, the active role of both UN Educational, Scientific and Cultural Organization (UNESCO) and the UN Development Programme (UNDP) in revitalizing and preserving historical buildings in downtown Lima, Peru, and providing housing for poor Peruvians living in that part of the city will also constrain any community development project attempted in that area. (See http://content.undp.org/go/newsroom/2009/december/per-preservar-el-patrimonio-histrico-y-evitar-desastres.en).

- Over the decades, *NGOs and relief organizations* have become important actors in SCD, especially in areas where government presence and services to civil populations are tenuous. You have seen how in both case studies presented in this book (Chapters 6 and 7) NGOs play a central role in facilitating and continuing SCD projects.

- *Government agencies* such as USAID and Peace Corps (US), GTZ (Germany) and Agence Française de Développement (AFD, France), just to name a few, will likely be present through funding and technical personnel in the communities where you want to do SCD work. Although this is by no means a comprehensive list, we just want to highlight that the size and kind of institutions on the ground, and the relationship among them and with local and national governments, play a significant role in the development and implementation of SCD projects.

Actors. Context and institutions will determine to a large extend who you might find on the ground to assist (or resist) you in your SCD efforts. Even within the same organization, you will find a diversity of actors such as anthropologists, rural sociologists, economists, scientists, technicians, nurses, doctors, etc. You will also find a complex diversity of actors within a community. Women, children, elderly, clans, traditional families, kinship groups, etc., would likely want different things from an SCD project; hence, each would likely define the problem and propose solutions differently. Understanding, respecting, and valuing the role of each of these perspectives, through contextual listening, for example, becomes important in the success of SCD projects.

We do not expect you to become an expert on each one of these elements of SCD. Yet if you ignore them, it would be at your own peril—and that of the communities you intend to serve. Hence, we invite you to consider a more complex picture of what SCD looks like in Figure 9.1.

As the arrows indicate, these institutions and stakeholders are interconnected in complex ways. The network of institutions and stakeholders is multifold, dynamic, and shifts power and resource constraints and opportunities for the community. Also, the network can be mutually shaping and interdependent, wherein a change in policy or assumptions in one institution, or a change in context, can have ripple effects across multiple components of the network.

9.2   RECOMMENDATIONS

By reading this book, completing the exercises throughout its chapters, and beginning to apply its lessons in your SCD-related activities, you have come a long way in becoming an engineer who can work with communities. But your journey into the world of SCD is just beginning. As this chapter briefly outlines, the world of SCD is complex. To become a more knowledgeable, responsible and caring actor who respects community, first and foremost, and understands the complex world of SCD, consider the following:

- Complement your engineering education and develop a life-long learning attitude by taking courses related to SCD (e.g., development studies, cultural anthropology, and international political economy) that will help you further understand, appreciate, and deal with the context, institutions, and actors that make the world of SCD.

- If you are committed to a career in SCD, embark in a graduate program related to SCD such as the Engineering for Developing Communities Program at University of Colorado-Boulder, Peace Corps Master's International Program in Civil and Environmental Engineering at Michigan Tech, the Masters Program in Humanitarian Assistance at Tufts University's Fe-

Figure 9.1: Network of interrelationships among principal stakeholders in SCD contexts. Note engineers' location in every kind of institution involved in SCD and expected collaborations (and conflicts) with multiple non-engineers actors.

instein International Center, or the Master's in Development Practice at Columbia University's Earth Institute.

- Intern or co-op, even as an unpaid volunteer, with SCD-related institutions such as Water for People (WFP), International Development Enterprises (IDE), or Mercy Corps. Jobs in corporate engineering are no substitute for experience in SCD-related jobs.

- Develop and enhance your ability to listen beyond basic listening by practicing contextual listening. To do so, one does not necessarily need to travel abroad. Such listening can be practiced by placing oneself in many unfamiliar contexts and beginning with a few questions: What beliefs and assumptions do I bring to this group? What kind of listening would help me understand how to see the world though the eyes of someone else?

- Lastly, pay great attention to the principles of peace —equity, social justice and reconciliation— since without peace there is no development of any kind.

As researchers and engineering educators, the act of researching and writing this book has led us to new questions. We move now to understand the relationship between engineering and social justice. As engineers today might be enacting various forms of social justice in SCD-related projects and programs, in the next couple of years we will try to answer the following questions:

- How are engineering students and faculty interpreting social justice?

- How do those interpretations intersect with their education and practices as engineers?

- What might engineering and social justice have in common?

- In which ways have these two fields aligned, clashed, or interfaced throughout recent US history?

- How are engineering and social justice practiced today?

We hope you join us in this next quest!

REFERENCES

Adas, M. (1989). *Machines as the Measure of Men: Science, Technology, and Ideologies of Western Dominance*. Ithaca, Cornell University Press. 204

Architecture for Humanity (2006). *Design like you give a damn: Architectural responses to humanitarian crises*. New York, Metropolis Books. 205

Baillie, C., E. Feinblatt, T. Thamae, and E. Berrington (2010). *Needs and Feasibility: A Guide for Engineers in Community Projects. The Case of Waste for Life*. San Rafael, CA, Morgan & Claypool. 206

Burkey, S. (1993). *People first: A guide to self-reliant participatory rural development*. London and New York, Zed Books. 205

Cooper-Hewitt National Design Museum (2007). *Design for the other 90%*. New York, Cooper-Hewitt National Design Museum. 205

Cuny, F. (1983). *Disasters and Development*. New York and Oxford, Oxford University Press. 206

Cuny, F. C. and R. B. Hill (1999). *Famine, conflict, and response: a basic guide*. West Hartford, Kumarian Press. 206

Easterly, W. (2006). *The White Man's Burden: Why the West's Efforts to Aid the Rest Have Done so Much Ill and so Little Good*. New York, The Penguin Press. 205

Kidder, T. (2004). *Mountains Beyond Mountains: The Quest of Dr. Paul Farmer, A Man Who Would Cure the World*. New York, Random House. 205, 206

Marx, L. (1987). "Does Improved Technology Mean Progress?" *Technology Review*: 33–41. 204

Nieusma, D. and D. Riley (2010). "Designs on Development: Engineering, Globalization, and Social Justice." *Engineering Studies* **2**(1). 207

Rambaldi, G., R. Chambers, M. McCall and J. Fox (2006). "Practical ethics for PGIS practitioners, facilitators, technology intermediaries and researchers." *Participatory learning and action*. 206

Riley, D. (2007). *Resisting neoliberalism in global development engineering*. 2007 ASEE Annual Conference and Exhibition, Hawaii, ASEE. 207

Riley, D. (2008). *Engineering and social justice*, Morgan & Claypool. 207

Salmen, L. F. and E. Kane (2006). *Bridging diversity: Participatory learning for responsive development*. Washington, D.C., The World Bank. 205

Scott, J. C. (1998). *Seeing like a state: how certain schemes to improve the human condition have failed*. New Haven, Yale University Press. 204

Slim, H. and P. Thomson (1995). *Listening for a change: Oral testimony and community development*. Philadelphia, PA, New Society Publishers. 205

Authors' Biographies

JUAN LUCENA

Juan Lucena is Associate Professor in the Liberal Arts and International Studies Division (LAIS) at the Colorado School of Mines, where he teaches courses for engineers in the Humanitarian Engineering minor. Juan obtained a Ph.D. in Science and Technology Studies (STS) from Virginia Tech and a MS in STS and BS in Mechanical and Aeronautical Engineering from Rensselaer Polytechnic Institute (RPI). He is the author of *Defending the Nation: U.S. Policymaking to Create Scientists and Engineers from Sputnik to the "War Against Terrorism"* (University Press of America, 2005). He has been principal investigator on research grants related to globalization and engineering, the national and cultural dimensions of engineering, humanitarianism and engineering, and engineering and social justice. Having served in key advising groups in engineering education and policy, he is currently a member of the advisory committee for the Center for Engineering Ethics and Society at the National Academy of Engineering. Juan is co-editor of *Engineering Studies,* the Journal of the International Network for Engineering Studies.

Email: jlucena@mines.edu

JEN SCHNEIDER

Jen Schneider is Assistant Professor in the Liberal Arts and International Studies Division (LAIS) at the Colorado School of Mines (CSM), where she has taught for seven years. Jen's Ph.D. and M.A. degrees are in Cultural Studies from Claremont Graduate University, where her work focused on popular cultural representations of "grotesque bodies" in the postwar era through a study of fiction and film. Since coming to CSM in 2003, Jen has applied ways to apply her education in critical theory and media studies to problems in science, engineering, and engineering education. Her current research interests address how scientists and engineers communicate with the public and media about emerging risks and environmental crises, such as climate change, nanotechnology, and nuclear power. She also analyzes popular media texts such as film to understand how the public, and engineers and scientists, develop mental and affective models related to these risks. At CSM, she teaches courses in communication and media studies. A second significant focus of Jen's work has to do with understanding how engineers work to address problems in environment and resources, primarily through projects in engineering, sustainable community development, and social justice. She co-teaches courses in engineering and sustainable community development (ESCD) with Jon, Juan, and others at CSM.

Email: jjschnei@mines.edu

JON A. LEYDENS

Jon A. Leydens is Associate Professor in the Division of Liberal Arts and International Studies (LAIS) at the Colorado School of Mines (CSM), where he has served as Writing Program Administrator since 1997. Jon obtained a Ph.D. in Education with a focus on the teaching and learning of writing from Colorado State University (CSU), where he also obtained his MA is in Composition and Rhetoric (English). His dissertation looked at the role of rhetoric in engineering workplaces. Since 1989, Jon has taught at the college level in Europe and the United States. He has served as the chair of the CSM Writing Across the Curriculum Committee since its inception in 1998, and since that year has held summer workshops for science and engineering faculty on writing in the disciplines. In addition to writing across the curriculum, his research interests include teaching, learning, and social justice issues. Besides a graduate course in academic publishing, he also teaches undergraduate courses in proposal writing (a service learning course), rhetoric, and mass media. He is also a co-principle investigator (PI) on a National Science Foundation grant to explore intersections between engineering and social justice as fields of practice. In 2008 and 2009, Jon co-designed and co-taught a seminar in Engineering and Sustainable Community Development.
Email: jleydens@mines.edu

Index

Made in the USA
San Bernardino, CA
30 November 2016